AF781973

BUSINESS ISSUES, COMPETITION AND ENTREPRENEURSHIP

FOREIGN-BORN STEM ENTREPRENEURSHIP

PARTICIPATION AND DIFFERENCES WITH U.S.-BORN WORKERS

BUSINESS ISSUES, COMPETITION AND ENTREPRENEURSHIP

Additional books in this series can be found on Nova's website under the Series tab.

Additional e-books in this series can be found on Nova's website under the eBooks tab.

BUSINESS ISSUES, COMPETITION AND ENTREPRENEURSHIP

FOREIGN-BORN STEM ENTREPRENEURSHIP

PARTICIPATION AND DIFFERENCES WITH U.S.-BORN WORKERS

WAYNE STEVENS
EDITOR

New York

Copyright © 2016 by Nova Science Publishers, Inc.

All rights reserved. No part of this book may be reproduced, stored in a retrieval system or transmitted in any form or by any means: electronic, electrostatic, magnetic, tape, mechanical photocopying, recording or otherwise without the written permission of the Publisher.

We have partnered with Copyright Clearance Center to make it easy for you to obtain permissions to reuse content from this publication. Simply navigate to this publication's page on Nova's website and locate the "Get Permission" button below the title description. This button is linked directly to the title's permission page on copyright.com. Alternatively, you can visit copyright.com and search by title, ISBN, or ISSN.

For further questions about using the service on copyright.com, please contact:
Copyright Clearance Center
Phone: +1-(978) 750-8400 Fax: +1-(978) 750-4470 E-mail: info@copyright.com.

NOTICE TO THE READER

The Publisher has taken reasonable care in the preparation of this book, but makes no expressed or implied warranty of any kind and assumes no responsibility for any errors or omissions. No liability is assumed for incidental or consequential damages in connection with or arising out of information contained in this book. The Publisher shall not be liable for any special, consequential, or exemplary damages resulting, in whole or in part, from the readers' use of, or reliance upon, this material. Any parts of this book based on government reports are so indicated and copyright is claimed for those parts to the extent applicable to compilations of such works.

Independent verification should be sought for any data, advice or recommendations contained in this book. In addition, no responsibility is assumed by the publisher for any injury and/or damage to persons or property arising from any methods, products, instructions, ideas or otherwise contained in this publication.

This publication is designed to provide accurate and authoritative information with regard to the subject matter covered herein. It is sold with the clear understanding that the Publisher is not engaged in rendering legal or any other professional services. If legal or any other expert assistance is required, the services of a competent person should be sought. FROM A DECLARATION OF PARTICIPANTS JOINTLY ADOPTED BY A COMMITTEE OF THE AMERICAN BAR ASSOCIATION AND A COMMITTEE OF PUBLISHERS.

Additional color graphics may be available in the e-book version of this book.

Library of Congress Cataloging-in-Publication Data

ISBN: 978-1-53610-270-3

Published by Nova Science Publishers, Inc. † New York

CONTENTS

PREFACE

The STEM fields—science, technology, engineering and mathematics—are the source of tangible innovations in products and processes that help to spur economic growth. Though many of these advances may occur in established organizations, radical innovation has long been associated with entrepreneurial ventures. Several previous studies have shown that high-growth, high-tech STEM-based businesses in the United States are disproportionately founded by foreign-born scientists and engineers. However, recent data also suggest that immigrants' rate of participation in U.S. entrepreneurship is slowing. Policies that support nascent immigrant STEM entrepreneurs may also help to improve U.S. employment rates, economic productivity, and career satisfaction among new Americans and legal permanent residents. This book investigates several explanations for differences in STEM entrepreneurship between college-educated native-born and foreign-born workers. It also explores reasons for differences in entrepreneurial participation among foreign-born workers.

In: Foreign-Born STEM Entrepreneurship ISBN: 978-1-53610-270-3
Editor: Wayne Stevens © 2016 Nova Science Publishers, Inc.

Chapter 1

IMPORTED ENTREPRENEURS: FOREIGN-BORN SCIENTISTS AND ENGINEERS IN U.S. STEM FIELDS ENTREPRENEURSHIP*

Margaret E. Blume-Kohout

This report was developed under a contract with the Small Business Administration, Office of Advocacy, and contains information and analysis that were reviewed by of the Office of Advocacy officials. However, the final conclusions of the report do not necessarily reflect the views of the Office of Advocacy.

EXECUTIVE SUMMARY

The STEM fields—science, technology, engineering and mathematics—are the source of tangible innovations in products and processes that help to spur economic growth. Though many of these advances may occur in established organizations, radical innovation has long been associated with entrepreneurial ventures. Several previous studies have shown that high-growth, high-tech

* This is an edited, reformatted and augmented version of a report developed under a contract with the Small Business Administration, Office of Advocacy and issued February 2016.

STEM-based businesses in the United States are disproportionately founded by foreign-born scientists and engineers. However, recent data also suggest that immigrants' rate of participation in U.S. entrepreneurship is slowing. Policies that support nascent immigrant STEM entrepreneurs may also help to improve U.S. employment rates, economic productivity, and career satisfaction among new Americans and legal permanent residents.

This report investigates several explanations for differences in STEM entrepreneurship between college-educated native-born and foreign-born workers. It also explores reasons for differences in entrepreneurial participation *among* foreign-born workers. These reasons may include:

- Immigration status and visa eligibility,
- Age of arrival in the U.S. (as a child versus as an adult),
- Graduating from a U.S. or foreign institution of higher learning, and
- Differences in cultural support for entrepreneurship across countries-of-origin.

We also describe differences in aspects of job satisfaction by citizenship and immigration status, providing insight into the preferences and work experiences of high-skilled foreign-born workers in the U.S. labor force.

Although we cannot identify with certainty the causes of recent observed declines in U.S. entrepreneurial participation among the foreign-born, our empirical results do suggest opportunities for policy and further research to help spur greater participation in the future.

Data

Analyses contained in this report primarily draw on recent nationally representative surveys of U.S.-resident workers holding bachelor's degrees and higher, combined in the National Science Foundation's 2010 Scientists and Engineers Statistical Data System (SESTAT) restricted-use data file. These data include information on the following:

- Foreign-born workers' current immigration status (i.e., naturalized U.S. citizen, legal permanent resident, or temporary resident),
- The U.S. state or foreign country where they graduated from high school,

- Where and in what field(s) they earned their first bachelor's and any subsequent higher education degrees, and
- Country of citizenship for foreign nationals.

Additional key variables include:

- Employment sector (e.g., academia, government, industry, non-profit organization, business owner, unincorporated self-employment, etc.),
- Employer size, and
- Demographic variables such as age, marital status, and gender.

However, one important limitation of these SESTAT data is that we cannot observe foreign temporary residents' specific visa type.

Key Findings and Policy Implications

- **U.S.–Foreign STEM Entrepreneurship Gap.** As in previous studies, we find a substantial percentage of the gap in rates of business ownership and STEM entrepreneurship among college-educated native U.S. and foreign-born workers can be attributed to differences in higher education attainment, particularly the undergraduate major they choose. Differences in the distribution of U.S. and foreign-born workers by age, experience, gender, and marital status explain most of the remaining gap, except among immigrants who earned their highest degree in the United States. For this group, the rate of STEM entrepreneurship is almost 4 percentage points higher than for otherwise-similar native U.S. citizens.
- **College-educated immigrants who earned their highest degree from a U.S. institution are more likely than native U.S. citizens and other foreign-born workers to engage in STEM entrepreneurship.** Intriguingly, whether these U.S.-trained adult immigrants earned graduate degrees in STEM versus non-STEM fields had no impact on their probability of STEM business ownership. Retaining more of these students in the United States post-graduation would tend to boost STEM entrepreneurship rates. Based on this report's empirical results, one way of achieving this goal would be to expand pathways for foreign temporary resident students who earn advanced degrees in the U.S. to

extend their stay after graduation, contingent on an offer of employment related to their U.S.-earned degree, regardless of employer sector. Successful examples of programs of this kind are described in the Policy and Conclusions section of the report.

- **STEM entrepreneurship rates among those who earned their highest degree abroad were significantly *lower* than one would otherwise expect.** This result is troubling because immigrants in this group who were employed by established organizations were also more likely to express dissatisfaction with their job's level of responsibility and intellectual challenge, indicative of skills-related underemployment. For some immigrants, occupational dissatisfaction and mismatch may spur a transition into entrepreneurship, which we find is associated with significantly greater satisfaction for adult immigrants, partly due to more appropriate use of their knowledge, skills, and abilities. For others, remaining in the U.S. may become increasingly unattractive, even after attaining legal permanent residence here. As long as so many highly skilled immigrants remain underemployed and underutilized, the U.S. will fall short of its potential for economic production and growth. Entrepreneurial venturing, training, and mentoring programs offered by higher education institutions might be particularly useful to help foreign-educated workers advance into STEM employment better suited to their skill level or into STEM entrepreneurship.
- **Foreign-born, foreign-educated workers (who earned STEM graduate degrees outside the U.S.) are more likely to work for startups than established organizations.** About half of firms that apply for H-1B visas on behalf of foreign temporary workers are small businesses with 30 or fewer employees (Doran, Gelber, and Isen 2015). Employment experience within a small firm can help individuals with entrepreneurial propensities to develop their entrepreneurial human capital, which in turn may promote better outcomes for their own future ventures (Elfenbein, Hamilton, and Zenger 2010). Programs that increase availability of highly skilled foreign graduates to new STEM ventures thus may be particularly helpful in spurring growth in U.S. innovative entrepreneurship, even if the foreign-born workers thus engaged are not themselves founders, owners, or investors.
- **Business ownership rates among foreign temporary resident workers from countries that are *not* eligible for E-2 Treaty Investor visas are lower than one would otherwise expect, based on those**

individuals' other characteristics. Relatively few options currently exist for foreign temporary workers to own and manage businesses in the United States. The most accessible, flexible option is the E-2 Treaty investor visa, which links U.S. residency to investment, management, and operation of a U.S. business. However, citizens of many countries—notably including India and China—are not eligible for E-2 visas; the absence of an equally accessible visa alternative for these foreign temporary resident workers appears to be stifling their participation as business owners in the U.S. economy. Nonetheless, despite the positive effect of E-2 Treaty country citizenship on foreign temporary residents' probability of business ownership, among STEM business owners we observe one outlier: over one-third (37%) of foreign temporary resident STEM business owners are from India. Overall, foreign-born founders of high-growth STEM ventures in the U.S. are significantly more likely to be of Indian origin, even after controlling for their other demographic and educational characteristics. More research is needed to understand which of the currently available visa mechanisms are most frequently utilized by foreign STEM entrepreneurs.

INTRODUCTION

Foreign-born scientists and engineers have contributed a great deal to U.S. economic growth, innovation, and prosperity. If upward mobility and equal opportunity for socioeconomic advancement are considered American values, then surely opportunity-driven immigrant entrepreneurs must endorse these values all the more strongly.[1]

In 2013, immigrants were almost twice as likely as native U.S. citizens to start businesses (Fairlee 2014). The Kauffman Index of Entrepreneurial Activity indicates immigrants' rate of entrepreneurial participation has been consistently higher than among U.S.-born since 1996, and this gap appears to have grown dramatically during the Great Recession. More than 1 in 4 (25.9%) new entrepreneurs in 2013 were immigrants, versus less than 1 in 5 (19.1%) a decade earlier (Fairlee 2014). Nationwide, about 1 in 4 engineering and technology-based companies are immigrant-founded, employing an estimated 560,000 workers and producing $63 billion in revenues (Wadwha, Saxenian, and Siciliano 2012). Although Fairlie (2008) found no significant differences in immigrants' and non-immigrants' distributions of entrepreneurship across industries—despite immigrants' typically higher levels of startup capital—several previous studies indicate that foreign-born workers are more likely than native U.S. citizens to start high-growth, high-tech companies. One recent nationally representative study estimated that 16% of rapidly growing high-impact, high-tech companies in the U.S. had at least one foreign-born founding member, and furthermore, companies that included immigrant entrepreneurs on their founding team performed better in some respects than similar companies with only U.S.-born founders (Hart and Ács 2011). Two other studies similarly estimated that 1 in 4 high-tech startups with more than $1 million in sales were founded by immigrants, and 1 in 3 founders of venture-backed companies that went public between 2006 and 2012 were immigrants (Anderson 2013; Wadwha, Saxenian, Freeman, and Gereffi 2007).

Despite these substantial contributions of foreign-born workers to U.S. economic growth, recent evidence suggests their participation in founding high-tech businesses may be slowing. In Silicon Valley, the share of businesses founded by foreign-born workers decreased from 52.4% to 43.9% from 2005 to 2012 (Wadwha, Saxenian, and Siciliano 2012).

These recent findings raise two tightly related questions. First, why do immigrants tend to have a higher rate of entrepreneurial participation than native U.S. citizens? Second, why do some groups of foreign-born workers seem to have higher propensity than others towards high-growth, high-impact, science-based entrepreneurship? Answers to these questions may help to inform immigration policy to better promote innovative entrepreneurship and economic growth.

Evidence from Prior Literature

Part of the answer to both of the questions posed above seems to reside in differences across groups in their typical educational preparation. In particular, much of the observed discrepancy between native U.S. citizens and immigrants in wages, innovative activity, and entrepreneurship can be explained by differences in their chosen fields of study (Hunt 2011; Hunt and Gauthier-Loiselle 2010). And yet, even after controlling for field of degree and highest level of education completed, Hunt (2011) found that college-educated immigrants who come to the U.S. for work or study are significantly more likely than native-born U.S. citizens to file for patents and start new companies, and she concluded there exists "a niche for immigrants in founding firms using specialized academic knowledge."

Another common explanation for higher entrepreneurial participation among the foreign-born asserts higher inherent quality or ability among those who come to the U.S., particularly with regard to science, technology, engineering and mathematics ("STEM") fields. Early work by Stephan and Levin (2001) documented immigrants' exceptional contributions as "superstars" in STEM fields, finding that immigrant scientists are disproportionately among the most highly cited authors, and are also more often elected to the National Academy of Sciences. While Kerr (2013) observes that immigrants' disproportionate presence in the high end of the distribution for scientific productivity does not necessarily imply higher *average* quality or innovativeness, increasing the share of immigrants has been shown to increase U.S. patenting and economic productivity (Hunt and Gauthier-Loiselle 2010; Peri, Shih, and Sparber 2015; Kerr and Lincoln 2010). Foreign-born faculty members also tend to produce more peer-reviewed publications than their native U.S. citizen colleagues, though this may be due to their typically spending a greater share of work time on research than on teaching or service (Webber 2012; No and Walsh 2010). No and Walsh's (2010) survey also showed slightly higher subsequent citations for patents with foreign-born lead inventors, after controlling for field, organization type, and so on.

The importance of workers' higher education attainment and STEM training in potentially spurring innovation and entrepreneurship can be further nuanced by asking, does it make a difference whether a foreign-born worker pursued higher education degrees in the U.S. or abroad? The increasing penetration of foreign students in U.S. graduate programs suggests potential for a growing role for U.S.-trained immigrants in the innovative, entrepreneurial workforce. Patents with U.S.-educated foreign-born lead inventors are significantly more likely to be commercialized (No and Walsh 2010). Just over half of Indian, Chinese and Taiwanese founders of U.S. engineering and technology firms initially came to the U.S. to pursue higher education, and earned their highest degree from a U.S. institution (Wadwha, Rissing, Saxenian, and Gereffi 2007). Evidence also suggests the increasing share of foreign students in U.S. graduate programs has itself increased U.S. innovation and productivity. Chellaraj et al. (2008) found that increasing the share of foreign graduate students significantly increased U.S. patent applications and awards, both for academic and non-academic institutions. If foreign students who pursue graduate degrees at U.S. universities are also more likely subsequently to engage in high-growth and innovative entrepreneurial ventures, that could provide support for immigration policy reforms expanding opportunities for foreign students who graduate from U.S. universities to stay in the U.S. and work after graduation. On the other hand, evidence from prior theoretical and empirical studies suggests that employers' lack of information about the quality of foreign higher education institutions (and labor market discrimination, more broadly) may result in employment mismatch, spurring high-ability immigrants trained abroad into self-employment (Åstebro, Chen, and Thompson 2011; Hegde and Tumlinson 2015).

Despite these previous findings, visa programs for foreign students and foreign temporary workers remain controversial. For example, some studies have found that an increasing share of foreign undergraduates decreases enrollment of native U.S. women STEM majors and crowds out non-Hispanic native white males in graduate programs (Orrenius and Zavodny 2015; Borjas 2007; Hoxby 1998). In theory, highly educated immigrants could also reduce wages and employment among similarly educated native U.S. workers (Stephan and Levin 2007; Freeman 2006; Borjas 2007). However, empirical evidence demonstrates that most of the growing demand for U.S. STEM workers over the past two decades has been met by immigrants and foreign temporary workers, with little to no effect on employment of similarly educated native U.S. workers (Kerr and Lincoln 2010; Peri and Sparber 2011; Peri, Shih, and Sparber 2015). Moreover, at least for lower-skilled immigrants, an increase in their labor supply has no effect on existing firms' total employment or wages, but rather increases the number of new establishments in the region (Olney 2013). If the same effect applies for highly educated foreign-born workers, we would expect to observe two related results: more new startups and higher rates of foreign-born workers joining new startup ventures.

The increasing attractiveness of other countries for foreign-born scientists and engineers combined with immigrants' slowing rate of entrepreneurial participation have raised concerns about the United States' ability to attract and retain innovative and entrepreneurial STEM workers. Recent studies indicate that foreign students from India and China increasingly are finding attractive job opportunities in their home countries, and visa considerations rarely contributed to returnees' decisions to go home (Wadwha, Saxenian, Freeman, and Gereffi 2007). Although many foreign-born STEM entrepreneurs indicate that the scientific infrastructure in their field was an important factor in their decision to come to the U.S., those who train in the U.S. then move to another high-income country with high scientific output can be equally productive in terms of scientific publications as if they had stayed in the U.S. (Kahn and MacGarvie 2014).

Though some of the decision to immigrate to the U.S. may be motivated by differences across countries in macroeconomic conditions and financial returns to higher education, we posit that differences across countries in sociocultural, economic, and regulatory support for entrepreneurship may be particularly important among foreign-born workers with entrepreneurial propensities.

For example, all else equal, citizens of foreign countries ineligible for E-2 Treaty Investor visas may be less likely to consider immigrating to the U.S. to pursue entrepreneurial ventures, because the visa mechanisms available to them impose substantially greater restrictions.[2] If we find that citizens of E-2 Treaty Investor countries are more likely to engage in entrepreneurship, this may suggest potential for an alternative but similarly flexible visa to encourage entrepreneurial foreign citizens from non-E-2 countries to come to the U.S. and start businesses.

In addition, from a sociocultural perspective, citizens of countries that offer as much (or more) moral and tangible support to highly educated entrepreneurs as compared to that found in the U.S. may have relatively little incentive to immigrate to the U.S. for the purpose of pursuing entrepreneurial ventures. Understanding such motivations among current foreign-born entrepreneurs may help us both to predict future trends in U.S. entrepreneurship and to identify characteristics and competitive practices of countries most attractive to innovative entrepreneurs.

This report examines several potential contributors to STEM entrepreneurship among U.S. resident workers with college degrees (bachelor's and higher). It compares probability of owning a business or joining a startup as an employee among foreign- and native-born U.S. workers with otherwise similar educational attainment and demographic characteristics. We employ data from recent nationally representative surveys of U.S. residents with bachelor's and higher degrees to test the importance of various correlates of entrepreneurial participation, including:

- Age of immigrants' arrival in the United States (as children or adults);
- Country granting immigrants' highest degree (U.S. university versus abroad);
- Foreign temporary resident workers' citizenship with an E-2 Treaty country;
- Educational attainment, including bachelor's field of degree, highest degree attained, and whether highest degree was also in a STEM field;
- Postgraduate years of experience;
- Country-of-origin support for entrepreneurship; and
- Demographic characteristics (age, gender, marital status).

The report is structured as follows. First, we review several visa options available for prospective foreign-born workers, the associated ease of engaging in entrepreneurship across these various mechanisms, as well as policy considerations studied in the academic literature. In the second section, we introduce the survey datasets used in our analyses, and compare rates of STEM employment and higher education attainment for native U.S. and foreign-born workers. Next, we present results from multivariate logistic regression models predicting entrepreneurial outcomes as a function of individual and country-of-origin characteristics. Then, we examine measures of job satisfaction among foreign-born STEM workers in the U.S., to identify whether disproportionate dissatisfaction with job opportunities—for example, underemployment vis-à-vis their skills—may be contributing to foreign-born workers' disenchantment. The final section concludes with a summary of our results and implications for policy alternatives.

Visa Options for Foreign Entrepreneurs in the United States

Many factors may influence an individual's decision to emigrate, including greater relative or absolute returns to higher education, initial financial wealth, language and cultural ties, geographic distance and cost of migration (Belot and Hatton 2012; Brücker and Defoort 2009; Borjas 1987; Clark, Hatton, and Williamson 2007). These incentives and constraints are further shaped by the immigration policies of prospective destination countries. This section reviews the principal visas available to enable foreign nationals' participation in entrepreneurship in the United States.[3] Table 1 summarizes the temporary resident visa mechanisms, and Table 2 covers those that confer permanent residence. Each visa category has numerous restrictions, advantages, and disadvantages. In addition, the legal and regulatory environment surrounding their use is complex and changing. As of February 2016, the U.S. Department of Homeland Security has released proposed rules and guidance documents that may update some of the visa categories discussed in this report; these proposals are referenced in footnotes.

Table 1. Selected U.S. Temporary Resident Visas with Potential for Use by Foreign-Born Entrepreneurs

Visa	Eligibility Requirements	Options for Entrepreneurs	Maximum Time Limit
F-1* Academic Student	• Acceptance by a certified U.S. educational institution • Full-time enrollment in approved course of study • Proof of ability to pay for tuition and living expenses • Declared intent to depart upon completion of program	Students who qualify for Optional Practical Training (OPT) employment authorization can work full-time in a business directly related to their program of study for up to 12 months following its completion. STEM graduates can apply for an additional 17-month extension.	Following graduation, OPT program participants in STEM fields may remain for 29 months total; 12 months for all others.
H-1B* Specialty Occupation	• Bachelor's or higher degree from U.S. or foreign institution • Working in an occupation that normally requires at least a bachelor's degree in business or a STEM field • Job is directly related to their field of degree • Employer petEIEhE1	An exception to the employer petition exists; entrepreneurs who provide evidence of an incorporated venture with a separate board of directors that has ability to fire, supervise, or otherwise control their employment may be eligible. Businesses are subject to annual cap on total H-1B visas issued.	Six years; initial visa is issued for three-year period, but is renewable.

Table 1. (Continued)

Visa	Eligibility Requirements	Options for Entrepreneurs	Maximum Time Limit
E-2* Treaty Investor	• Substantial and irrevocable investment in a business majority-owned by citizens of the applicant's country • Home country is on the State Department E-2 Treaty Country list.	Executive leadership team members and essential employees with highly specialized skills are eligible, provided the applicant can invest sufficient funds. Visa requires beneficiary to operate, manage, or be employed by the organization.	Indefinite; initial visa is issued for two years, but is renewable as long as business continues to operate and creates jobs for U.S. workers or other positive local impacts.
L-1A* Intra-company Transferee	• Foreign company executives or managers, entering the U.S. to establish a U.S. office • Foreign company must have existed for at least one year, must continue to exist and do business abroad, and must share common ownership and control with the new U.S. business	Entrepreneur must have worked abroad for at least one continuous year within the past three years, and in a qualifying position with the foreign entity. Visa requires beneficiary to manage and be employed by the organization.	Seven years total; initial visa is issued for one year, but can be renewed in two-year increments if office is active and operating, with physical premises, hired employees, and sufficient revenues or investment to support the L-1 visa holder(s).

*DHS released a proposed rule revising the F-1 visa program in October 2015, and a proposed rule updating the EB-1, EB-2, EB-3 and H-1B programs in December 2015. As of February 2016, these rules had not been finalized.

Source: "Entrepreneur Visa Guide," U.S. Citizenship and Immigration Service. Available at: www.uscis.gov/eir/visa-guide/entrepreneur-visa-guide, last accessed July 24, 2015.

F-1 Academic Student Visas

The U.S. attracts more foreign students than any other country, though several countries including the United Kingdom, France, Australia, Germany, and Canada have a greater share of foreign students enrolled in their higher education programs (Suter and Jandl 2008). Foreign citizens wishing to pursue higher education degrees in the United States generally will obtain an F-1 Academic Student non-immigrant visa, which allows them to enroll full-time in eligible U.S. degree- and certificate-granting programs. From 2008 to 2012, over 30% of foreign undergraduate and graduate students on F-1 visas were studying business, management, or marketing. By contrast, approximately 60% of foreign PhD students and 37% of foreign students, overall, were pursuing STEM degrees (Ruiz 2014).

F-1 visa holders are permitted to work on campus or in school-approved off-campus positions through the Curricular Practical Training (CPT) program, provided their employment is directly related to their degree program and field of study. After completing their degree, students on F-1 visas are generally eligible to apply for Optional Practical Training (OPT), which permits them to remain in the U.S. for 12 months of full-time work for each postsecondary degree, plus an additional 17 months if their degree was in a STEM field.

Starting a business is generally permitted while on F-1 OPT, provided that the majority of work performed is directly related to their major field of degree.[4] However, it appears to be rarely used for this purpose. Data from the 2010 National Survey of College Graduates indicate that less than 4% of college-educated foreign temporary resident STEM business owners currently hold a student or trainee visa.[5]

The number of foreign students enrolling in U.S. universities through the F-1 visa program has increased markedly over the past few decades. Prospective foreign students may be attracted to the U.S. by the increased odds of achieving permanent residence following their U.S.-earned degree. Borjas (2002) estimated the probability of ultimately becoming a permanent resident was 26 times higher for those who earned degrees in the U.S., versus through the random "green card" lottery. Relatedly, Kato and Sparber (2013) conversely found that restricting availability of H-1B temporary worker visas reduces the number of high-quality, foreign prospective students applying to U.S. undergraduate programs.

Despite the apparent benefits to U.S. innovation and economic productivity, higher education investments in foreign students remain controversial. Many foreign students who pursue advanced degrees in the U.S. do remain in the U.S. workforce after graduation, and in addition to enhancing universities' research productivity and curricular diversity, those who pay tuition with funding from non-U.S. sources may help to support public goods. But, substantial differences in foreign students' stay rates exist across degree fields and countries of citizenship (Finn 2010). As Winston (1999) discusses, graduate education in particular benefits from public subsidies, and this trend has raised concerns that U.S. taxpayer funds are subsidizing foreign nations' competitiveness (Borjas 2007; Winston 1999; Borjas 2002). Finn (2010) estimates 75% of foreign students who earned PhDs in computer science and electrical engineering in 2002 were still living and working in the U.S. five years later, versus less than half of those who earned PhDs in agricultural sciences.

To some extent, these field-specific differences in foreign students' stay rates reflect differences across sending countries in the U.S. degrees that students from each country tend to pursue. Saxenian (1999) observed that "Chinese and Indian engineers, most of whom arrived in the United States after 1970 to pursue graduate studies, were senior executives at one-quarter of Silicon Valley's new technology businesses," contributing over $16.8 billion in sales and over 58,000 jobs. Five-year stay rates are highest among Chinese, Indian, and Eastern European students, with over 80% of STEM PhDs from these countries still in the U.S. five years after attaining their doctorates (Finn 2010). Students from these countries are also heavily represented among U.S.-earned PhDs in computer science, electrical engineering, and in the physical sciences, but they comprise a markedly lower share of foreign students earning PhDs in agricultural sciences, economics, and other social sciences.

Hsu et al. (2007) suggest that foreign students who pursue higher education in the U.S. tend to be especially wealthy, relative to most in their home countries. Greater financial wealth enables families to pay upfront migration and transportation costs, but also may have enabled those families to provide superior primary and secondary educational opportunities at home, better preparing candidates for admission to U.S. universities. U.S. immigration policy also requires student visa applicants to demonstrate *a priori* adequate financial resources to support themselves and any dependents for the length of their stay, which may further reinforce this wealth effect. For some foreign students who wish to stay in the U.S. after graduation, nonimmigrant E-2 Treaty Investor visas thus might be a feasible solution, particularly for students who were able to afford out-of-pocket the cost of attending elite research universities (Hsu, Roberts, and Eesley 2007). Such selection on wealth would also appear consistent with Fairlee's (2012) finding that immigrant-owned businesses tend to have higher levels of financial capital at startup, compared to businesses founded by U.S. natives, with a greater share of startup funds derived from personal and family sources.

On the other hand, Stuen et al. (2012) find that increases in foreign graduate students who pay their own tuition yield significantly less benefit for academic patenting, as compared to increases in foreign students whose quality earned them fellowships, assistantships, or other types of funding support. Stuen et al. therefore recommend student visa decisions be made on the basis of admission offers from top universities, rather than requiring evidence of substantial financial resources. Others have argued that foreign students completing U.S. degrees in computer science or electrical engineering, and who subsequently obtain H-1B visas to work in IT industry jobs, tend to graduate from lower-ranked programs (Matloff 2013). Further research is needed to understand the extent to which students at these lower-ranked programs are selected on academic quality versus ability to pay, as well as their subsequent contributions to patenting, innovation, and economic growth. Finally, although some complementary descriptive evidence also suggests graduates from top-ranked programs may be less likely to stay in the U.S. after graduation, this may simply reflect differences across sending countries in probability of enrolling in higher-ranked programs, and differences in the probability that graduates from each country will stay (Finn 2010).

Grogger and Hanson (2015) found that, overall, foreign students awarded U.S.-based fellowships, research assistantships, and teaching assistantships are more likely to intend to stay in the U.S. after graduation. Their finding may partly be due to quality effects of the institution attended, as higher-ranked programs more often provide full support for admitted students. Controlling for these differences across institutions, Blume-Kohout and Adhikari (2014) find no significant difference in short-run stay rates or U.S. scientific workforce participation among foreign students whose primary support came from university assistantships versus personal or family funds, controlling for university attended and field of degree. On the other hand, foreign students who rely on other foreign sources of support are overwhelmingly less likely to stay, perhaps due to conditions of the financial assistance they receive (Blume-Kohout and Adhikari 2014; Grogger and Hanson 2015). Together, these results suggest there should be no detrimental effect on foreign students' stay rates if visas are extended to prospective graduate students who are offered admission and funded assistantships from U.S. universities, independent of their other personal or family financial resources.

H-1B Specialty Occupation

The H-1B Specialty Occupation visa is available to foreign citizens who hold bachelor's and higher degrees, and who will work in occupations that are both directly related to their field of degree and that would normally require at least a bachelor's degree in business or STEM fields. The H-1B is issued for a three-year period, and is renewable up to six years. As of 2013, over three-quarters (76%) of U.S. adults favored legislation to increase the number of visas available for temporary skilled workers (Dugan 2013).

Figure 1 depicts trends over time in F-1 and H-1B visa issuances. In 2010, approximately one-third of new H1-B temporary worker visas were awarded to foreign individuals transferring from F-1 student visas, of whom approximately three-quarters held graduate degrees (Ruiz 2014). Including both those trained in the U.S. and those who earned their highest degree abroad, nearly half (47%) of workers on H-1B visas in 2010 were in computing-related occupations, totaling over 90,000 (Wasem 2012). The substantially faster growth in F-1 visas versus H-1Bs in recent years implies that a smaller percentage of foreign students may be remaining in the U.S. after graduation. Our analysis of the recently released 2013 National Survey of College Graduates (NSCG) similarly indicates that two in five college-educated, U.S. resident foreign temporary workers are in computer, information technology, and electrical engineering occupations. This share is the same for those who are organizationally employed as well as for business owners, however, these business owners likely hold visas other than the H-1B.

The H-1B visa requires petition by a bona fide U.S. employer, and one would not expect entrepreneurial venturing among H-1B visa holders, as it effectively "locks" the foreign worker to the employer who petitioned on his or her behalf. Returns to scale in candidate search and the visa application process may also favor larger firms over startups as employers (Kerr, Kerr, and Lincoln 2015a). Foreign citizens wishing to engage in unincorporated self-employment are unlikely to be eligible, as there would be no separation between the petitioning employer and the beneficiary (USCIS 2010). However, under certain circumstances, the sole or majority owner of an incorporated entrepreneurial venture may be eligible for an H-1B visa. Specifically, "if the petitioner provides evidence that there is a separate Board of Directors which has the ability to hire, fire, pay, supervise or otherwise control the beneficiary's employment, the petitioner may be able to establish an employer-employee relationship with the beneficiary" (UCSIS 2010).

The number of H-1B visas issued annually is limited, though several exceptions exist. While for-profit employers are subject to an annual cap on H-1B visas, the first 20,000 applications filed on behalf of beneficiaries holding advanced degrees from U.S. universities are exempt from this cap. Similar temporary worker programs which are not subject to the H-1B cap exist for college-educated citizens of Chile and Singapore (H-1B1), Canada and Mexico (TN), and Australia (E-3). Importantly, U.S. colleges and universities, university-affiliated research institutes, and other not-for-profit research organizations are also exempt from the H-1B cap. This has given rise to another pathway for foreign citizens with advanced degrees to participate in entrepreneurship. In 2014, Massachusetts established the Global Entrepreneur in Residence (GEIR) program,[6] with two University of Massachusetts campuses as pilot sites. The GEIR program allows founders and leaders of startup ventures with advanced degrees in business or STEM fields to hold part-time employment with UMass-Boston or UMass-Lowell, while working to grow their Massachusetts-headquartered new businesses.

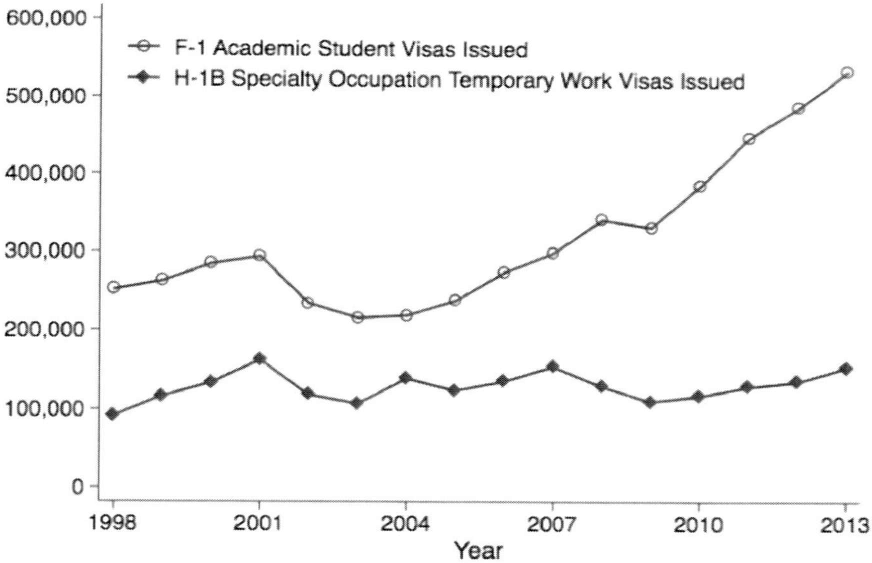

Source: U.S. Department of State, Nonimmigrant Visa Issuances by Visa Class.

Figure 1. Trends in F-1 Student Visas and H-1B Temporary Worker Visas, 1998-2013.

The H-1B program is also controversial. Foreign temporary workers overwhelmingly are hired into information technology jobs, and some studies suggest these workers displace trained U.S. graduates in these fields, in part by lowering the prevailing wages for computing-related occupations (Salzman, Kuehn, and Lowell 2013; Bound, Braga, Golden, and Khanna 2015). Recent careful work by Kerr and colleagues demonstrates that, within individual firms, increasing employment of immigrants in STEM occupations may also disproportionately displace older workers (Kerr, Kerr, and Lincoln 2015b).

One drawback of the current H-1B program which may affect its attractiveness to highly skilled workers and its use by foreign nationals launching entrepreneurial ventures is that spouses have not been eligible for derivative open work permits. A recent policy change now allows spouses to apply for work permits, but only after the H-1B holder is "in queue" with an approved petition for legal permanent residence.[7] This contrasts starkly with Canada's equivalent temporary skilled worker program, in which spouses can obtain open work permits, regardless of their own qualifications.

E-2 Treaty Investor

The E-2 Treaty Investor visa is available to citizens of specific countries with which the U.S. maintains treaties of commerce and navigation. It allows them to enter the U.S. for the purpose of starting or establishing a new commercial business, or purchasing and managing an existing commercial business such as a franchise, with majority ownership by citizens of the applicant's treaty country.[8] In fiscal year 2013, about 42% of E-2 Treaty Investor visas were issued to citizens of Asian countries, of which most (75%) were issued to Japanese investors. Another 38% were issued to Europeans, with German investors most common. Investors from the United Kingdom, Mexico, and Canada comprised 7% to 8% each.

The E-2 visa requires that substantial funds be actively and irrevocably invested by the applicant, towards the production of goods and services to be sold for profit. Because only the organization's executive leadership and essential employees with highly specialized skills are eligible for this visa category, simply having an ownership interest in a business is insufficient. The business must also generate employment or other economic impact above and beyond simply income support for the applicant and his or her immediate family members. Although the E-2 Treaty Investor visa is technically a non-immigrant visa, as long as the business continues to exist, satisfies the ownership requirement noted above, and either creates jobs for U.S. workers or generates other positive local economic impacts, the visa can be extended indefinitely beyond the initial two-year period. Despite its restricting eligibility to citizens of specific countries, the E-2 category is more broadly accessible than most other options for prospective immigrant entrepreneurs; it has no specific educational or work experience requirement, no numeric job-creation target, and substantially lower financial investment compared to the EB-5 visa, described later. In addition, spouses of E-2 Treaty investors are eligible to apply for work permits, and both the investor and his or her dependents up to age 21 can enroll in school without need for a separate visa. However, the list of E-2 Treaty countries notably excludes India and China, whose citizens together represent over half of college-educated business owners on foreign temporary worker visas.

L-1A, L-1B Intracompany Transferees[9]

Although L-1 visas are primarily used by U.S. companies wishing to bring a foreign national employed by a foreign subsidiary or office of their own organization, these visas can also be used by foreign companies that have been in existence for at least one year to transfer executives, managers, and other key personnel with specialized knowledge to the U.S., for the purpose of establishing a U.S. office. Prior literature has identified individuals in this latter group as "transnational entrepreneurs" (Drori, Honig, and Wright 2009). Approximately 28% of L-1 visas in FY2014 were issued to citizens of India.

The foreign company must demonstrate sufficient financial resources to compensate the L-1 employee(s) and to secure physical premises in the U.S., and it must continue to exist and do business abroad. Though initial stay for new U.S. offices is one year, the visa can be renewed in two-year increments, up to seven years for managers and executives. Unlike the E-2 Treaty Investor visa, L-1 visas are not restricted to a subset of countries. The L-1 visa may also be preferable for married workers, as the spouse of an L-1 worker (on an L-2 visa) can apply for an open work permit without limitations on their employment.

EB-5 Immigrant Investor

In contrast with the E-2 Treaty Investor non-immigrant visa, the EB-5 Immigrant Investor visa conveys permanent residence, i.e., a "green card," to an investor and his or her immediate family members, conditional on successful creation of at least 10 full-time jobs for U.S. workers within the initial two year period (Table 2). EB-5 investors must invest a minimum of $500,000 for businesses in designated rural or high-unemployment areas, and at least $1 million otherwise. Fewer than 10,000 EB-5 visas are awarded annually, approximately 65% of which are issued to investors' spouses and dependent children. Since 2008, Chinese citizens have claimed the largest share of EB-5 visas, and they received over 85% of the EB-5 visas issued in 2014 (U.S. Department of State 2014).

Table 2. Selected U.S. Permanent Resident Visas with Potential for Use by Foreign-Born Entrepreneurs

Visa	Eligibility Requirements	Options for Entrepreneurs	Maximum Time Limit
EB-5* Immigrant Investor	Minimum investment of $500,000 in a lawful commercial enterprise in designated rural or high unemployment areas, $1 million otherwise.	Visa does not require investor to have any active managerial or employment role with the business. Rarely used for investments in startup companies or projects in STEM-related industries.	None; Permanent resident visa issued following two- year trial period, conditional on creation of at least 10 full-time jobs for U.S. workers during that period.
EB-1, EB-2* Employment-Based Preference, Exceptional Ability	Foreign nationals with extraordinary ability or achievements in their field International recognition, advanced degrees, and 10 or more years of related work experience among the factors considered	EB-1–eligible individuals can apply on his or her own behalf to start a business related to their field of extraordinary ability or achievement. EB-2 eligible individuals must demonstrate U.S. national interest, e.g., holding licensed technologies or significant patents currently in use.	None; Permanent resident visa. (A temporary visa category also exists, the O-1 nonimmigrant visa, for individuals with extraordinary ability or achievement.).

*In December 2015, the Department of Homeland Security released a proposed rule updating the EB-1, EB-2, EB-3 and H-1B programs; this rule had not been finalized as of February 2016.

Source: "Entrepreneur Visa Guide," U.S. Citizenship and Immigration Service. Available at: http://www.uscis.gov/eir/visa-guide/entrepreneur-visa-guide, last accessed July 24, 2015.

Back-of-the-envelope calculations using U.S. Department of Homeland Security (DHS) statistics on EB-5 visas issued to investors and U.S. Citizenship and Immigration Service (USCIS) statistics for I-829 petition approvals[10] suggest that the success rate for EB-5 immigrant investor enterprises started between 2009 and 2011 was over 90%, substantially higher than the two-year survival rate for all U.S. businesses (approximately 70%) (BLS 2014). However, this comparison is misleading, due to the role of EB-5 Regional Centers in identifying and promoting large-scale economic development projects (e.g., commercial real estate, hotels, retailers, etc.) for prospective immigrants' direct investment, pooling capital from multiple investors, and providing model-based estimates of the projects' employment effects. Unlike the E-2 and H-1B visas which require the beneficiary to operate, manage, or be employed by a specific organization, EB-5 investors do not need to have any active role in the business in which they invest. The advocacy group Invest In the USA (IIUSA) estimates 95% of all EB-5 capital investment is raised through these Regional Centers, but very few of these Centers represent startup companies or projects in STEM-related industries. Thus, although in principle the EB-5 visa could be used by wealthy foreign entrepreneurs to establish innovative businesses in the United States, in practice this seems rarely to occur.

EB-1, EB-2, O-1: Extraordinary Ability and Achievement

The EB-1 immigrant visa and its counterpart, the temporary O-1 visa, are reserved for foreign nationals recognized for extraordinary achievements in their field of science, art, business, education, or athletics. Individuals who intend to start a business in the same field as their established extraordinary ability or achievements may qualify for these, and those wishing to immigrate can petition for an EB-1 visa on their own behalf. In addition, a U.S. organization owned by the beneficiary can petition on his or her behalf for an O-1 visa. In 2013, only 12,359 O-1 visas were issued, over half of them to citizens of Europe.

The level of international recognition and achievement required for these visas makes them generally inaccessible to recent graduates or workers with fewer years of traditional work experience. These limitations also apply to the EB-2 employment-based permanent resident visa. However, UCSIS recently clarified that the EB-2 visa requirement of a job offer from a U.S. employer could be waived if an entrepreneur can show their venture will have national impact and importance. Some examples of evidence that may support this national interest waiver for potential STEM entrepreneurs include documentation of having licensed a technology currently in use by others, or holding significant patents with documented usefulness to one's field.[11] Although not necessarily required, one factor considered in determining exceptional ability is having at least ten years full-time work experience in one's field.[12]

DATA

Data analyzed for this report include the National Science Foundation's (NSF) 2010 Scientists and Engineers Statistical Data System (SESTAT) restricted-use file, as well as country-level statistics derived from the Global Entrepreneurship Monitor (GEM) Adult Population Survey (APS), and the United Nations Educational, Scientific, and Cultural Organization (UNESCO) Institute for Statistics.

We use these data to explore the relative importance of each of the following possible factors affecting rates of business ownership and STEM entrepreneurship among foreign-born workers in the United States:

- Field of bachelor's degree;
- Highest educational attainment (bachelor's, master's, or research doctorate);
- Whether their highest degree was in a STEM field;
- Whether they hold an MBA or similar degree from a U.S. institution;
- Experience, measured as years since highest degree was awarded;
- Immigration status (naturalized U.S. citizen or legal permanent resident, versus on a foreign temporary resident visa);
- For naturalized U.S. citizens and legal permanent residents ("immigrants"), whether they came to the U.S. as children (i.e., before graduating from high school), or as adults;
- For adult immigrants, whether they came to the U.S. to pursue a higher education degree, and earned their highest or most recent degree here;
- For foreign temporary residents, whether they are from countries eligible for E-2 Treaty Investor visas;
- Demographic characteristics including age, gender, and marital status; and
- Country-of-origin characteristics, including cultural support for entrepreneurship, and share of population with postsecondary education.

Scientists and Engineers Statistical Data System (SESTAT)

The SESTAT variables used in this report are drawn from two component surveys: the National Survey of College Graduates (NSCG) and the Survey of Doctorate Recipients (SDR). Access to the restricted-use SESTAT microdata files was provided under NSF license, through NORC's Virtual Data Enclave.

The NSCG is the core SESTAT dataset, covering the entire population of U.S.-resident college graduates under age 76 who hold bachelor's or higher degrees.[13] Although in previous years the NSCG sampling frame included only those who either (a) were employed in STEM occupations, or (b) had received a bachelor's or higher degree in a STEM or STEM-related field, the 2010 NSCG sample was expanded to provide a representative sample of the U.S. resident college-educated workforce by including bachelor's-educated respondents in non-STEM fields from the 2009 American Community Survey sample.

The SDR is a longitudinal survey of U.S. residents who earned a research doctorate from a U.S. academic institution in natural sciences, engineering, health-related fields, or social/behavioral sciences. Sampled individuals are followed through their careers from the time their degree is awarded until age 76, with newly earned PhDs added in each survey cycle. This unique panel survey provides opportunities to examine responses to topical module questions of current interest for a large, nationally representative cross-sectional sample, and to follow a slightly smaller sample over time (due to individuals lost to follow-up). In the combined SESTAT file, survey weights are provided to accommodate over-sampling of PhDs due to the inclusion of these SDR respondents, as well as over-sampling of STEM workers overall in the NSCG, allowing nationally representative estimates. All empirical results presented in this report were calculated using Stata statistical software, applying these survey weights.

The NSCG and SDR ask a series of questions in each survey year regarding employment and job changes. The surveys also capture whether individuals are underemployed (working part-time when they would prefer full-time employment), and if they are not currently working, the reason(s) they give, for example inability to find a suitable job in their field. The surveys also capture any additional work-related training the respondents pursued, including whether they recently earned another degree, such as a master's in business administration (MBA). Finally, in 2010 the NSCG and SDR instruments both included an additional module focusing on aspects of individuals' job satisfaction. These additional variables permit us to compare, across citizenship and immigration status groups, the extent to which workers are satisfied with their pay, level of responsibility, degree of independence, and opportunities for advancement.

Defining STEM Entrepreneurship

Entrepreneurship is frequently defined in empirical research to cover both self-employment and business ownership. By this definition, the 2010 SESTAT data indicate that 18.2% of adults in the U.S. labor force who hold bachelor's or higher degrees are entrepreneurs, with relatively little difference between U.S. native and foreign-born workers. [14] About 18.8% of foreign-born and 18.0% of U.S. natives in the college-educated workforce participated in entrepreneurial activities 2010, and about 3.5% in each group were unincorporated self-employed, with no employees.[15]

However, the entrepreneurial gap between native U.S. citizens and foreign-born begins to widen when we consider business owners alone, and also if we account for the substantially higher rate of unemployment among the foreign-born. Among *employed* U.S. residents holding bachelor's or higher degrees, 16.8% of foreign-born workers identified as business owners or incorporated self-employed, versus 15.0% of U.S. natives with similar educational attainment.

For this study, as in Blume-Kohout (2014), we expand the definition of entrepreneurship further to include employees who join startup ventures (i.e., not just business owners). Conceptually, this follows Lazear's (2005) description of an entrepreneur as "among those who initially established the business." As Roach and Sauermann (2014) point out, nascent entrepreneurial "joiners" tend to share many of the entrepreneurial preferences typically ascribed to founding owners, such as desire for autonomy and tolerance for risk. Operationally, we define these "joiners" to include individuals who began work in 2005 or later with a U.S.-based private sector business or other non-governmental, non-academic organization, which had "come into being as a new business within the past five years," and which as of 2010 had fewer than 100 employees.

Figure 2 depicts percentages of the U.S. college-educated workforce who are self-employed, business owners, or participating as employees of new ventures. As Figure 2 shows, the total rates of entrepreneurial participation are similar for native U.S. citizens and immigrants who came to the U.S. as children and for higher education. Two categories are somewhat more likely than native U.S. citizens to participate in entrepreneurship: adult immigrants whose highest degree was earned outside of the U.S. and foreign-born workers on temporary resident visas from E-2 Treaty countries. By contrast, STEM entrepreneurship is most common among adult immigrants who came to the U.S. for higher education. Not surprisingly, given the restrictions on their work opportunities, foreign-born temporary resident workers from non-E-2 countries are the least likely to engage in entrepreneurship. These citizenship and immigration categories are described in more detail below, and in Table 3.

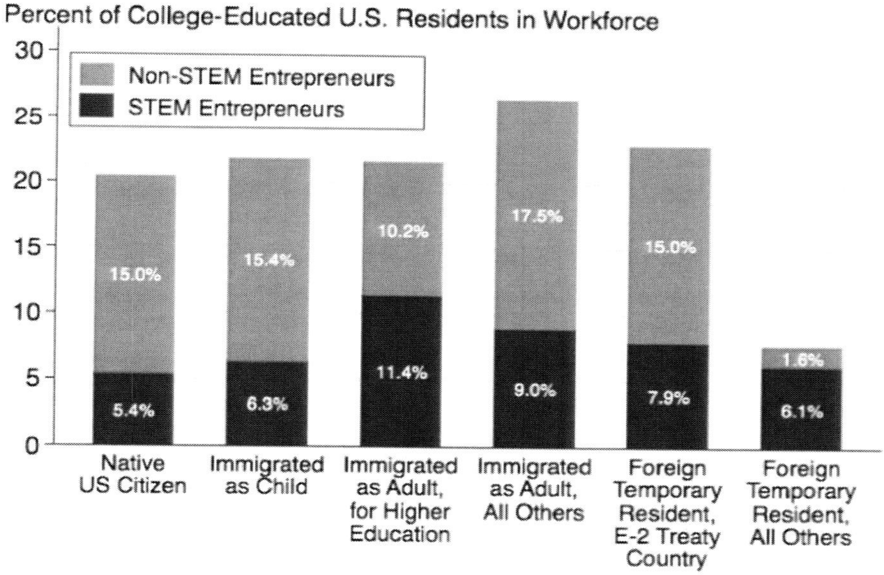

Source: Author's calculations, 2010 SESTAT data.

Figure 2. Entrepreneurship Participation by Citizenship and Immigration Status.

STEM entrepreneurs in Figure 2 are those self-employed, U.S. business owners, or new venture joiners who are also identified as working in STEM occupations. We define STEM occupations first by using the U.S. Bureau of Labor Statistics (2012) list, including all life scientists, physical scientists, engineers, and those in mathematics or statistical sciences occupations. To the specified natural sciences and engineering managers which BLS includes, we add other managerial or executive occupations (including, for example, "top-level executive") if the respondent answered affirmatively to the question: "Did your duties on this job require the technical expertise of a bachelor's degree or higher in engineering, computer science, math, or the natural sciences?" However, we exclude from STEM occupations any medical clinicians and allied health practitioners, clinical psychologists, and other BLS "health occupations," unless the respondent indicates that at least 10% of their work activities include basic research, applied research, development or design. For psychologists, we further require an affirmative response on the STEM technical expertise question. A similar approach combining STEM occupations and research-related work activities was recently used by Kahn et al. to identify scientific entrepreneurs in these data (Kahn, La Mattina, and MacGarvie 2015). Excluding traditional providers of clinical health services from our primary outcome variable allows greater focus on individuals most likely to contribute to high-growth innovative entrepreneurship.

Over three-quarters (75.7%) of all STEM entrepreneurs (per our definition) identified themselves as business owners, and about 1 in 8 (12.5%) identified as unincorporated self-employed, supervising one or more employees. The remaining 11.8% were "joiners"— employees in new startup ventures.

Overall, foreign-born entrepreneurs are significantly more likely than native U.S. citizens to own or join startup businesses requiring STEM fields expertise. About half (51%) of foreign-born entrepreneurs said their work required STEM technical knowledge, versus less than 40% of U.S. native entrepreneurs. However, as Hunt (2011) and others have noted, this distinction between U.S. natives and foreign-born masks dramatic differences in entrepreneurial participation across foreign-born workers, depending on the circumstances of their entry into the United States. Indeed, even U.S. citizens who lived outside the United States in their youth[16] are more likely than those raised in the U.S. to engage in entrepreneurship.

To better focus on entrepreneurs who contribute the most to economic growth and employment (and in contrast with Figure 2, which included all self-employed individuals as entrepreneurs), our econometric analysis in the next section excludes unincorporated self-employed individuals with no employees or direct reports (see footnote 15). Excluding this latter group, we estimate 19.0% of foreign-born and 16.5% of native U.S. citizen workers participated in entrepreneurship, for a total entrepreneurship gap of 2.5 percentage points (STEM and non-STEM).

Citizenship and Immigration Variables

Both the NSCG and the SDR include variables for type of citizenship (native or naturalized U.S. citizen, or foreign citizen), and for foreign citizens, whether they are on permanent or temporary resident visas. However, only the

NSCG collects more specific information about foreign temporary residents' visa type—e.g., worker, student, spouse/dependent, or other—at the time of the survey. Unfortunately, the survey does not ask for the specific type of visa, but rather presents a series of categorical responses with a few provided examples:

> (If a Temporary U.S. Resident Visa Holder)
> On October 1, 2010, did you hold a visa issued...
> 1. For temporary work (e.g., H-1B, L-1A, L-1B, etc.)
> 2. For study or training (e.g., F-1, J-1, H-3, etc.)
> 3. To you as the dependent of another person (e.g., F-2, H-4, J-2, K-2, L-2, etc.)
> 4. For any other reason

In addition to asking about current visa type, the NSCG includes a question about the type of visa the respondent had when he or she first came to the U.S. for a period of six months or longer, that is—whether as a permanent resident or one of the visa types listed previously. It is important to note, however, that this question does not specify the visa type at the individual's most recent long-term entry. This distinction has practical implications for these responses' use, as for example, individuals who first entered the U.S. as children of temporary workers but then returned to their home country, completed college and even graduate school abroad, may then return as adults on their own temporary worker visas, and ultimately acquire permanent residence or become naturalized citizens. Qualitatively and practically speaking, the unobserved abilities and preferences of an individual who immigrates to the U.S. as an adult worker but had some childhood experience of living in the U.S. may be significantly different from those of an individual who first entered the U.S. as, for example, the trailing H-2 spouse on their partner's H-1B visa. With this NSCG variable, however, we cannot distinguish between the two, as both are coded as having entered as dependents rather than for their own work or studies.

Due both to the absence of any similarly detailed question for the SDR respondents in the SESTAT data, and to the difficulties these NSCG responses present for categorizing "types" of immigrants, we combined NSCG responses (when available) with several other variables present in both datasets to better identify the circumstances of individuals' immigration to the U.S. and categorize immigrants in a policy-relevant way, such as:

- Country of citizenship, if not naturalized U.S. citizens
- Birthplace

Table 3. Summary of Citizenship and Immigrant Status Variables

Category	Description
Native U.S. Citizen	Individual born in the 50 U.S. states or U.S. territories, including Puerto Rico, or born abroad of American parents. Includes some individuals who identified themselves as naturalized citizens despite reporting a U.S. birthplace.
Foreign-Born: Immigrated as Child	Naturalized U.S. citizens and legal permanent residents who were born abroad, not of American parents, and who report they attended high school within the 50 U.S. states, and/or (for NSCG) they first entered the U.S. prior to age 18 on a dependent or student visa, or, most commonly, as permanent residents.
Foreign-Born: Immigrated as Adult, for Higher Education	Naturalized U.S. citizens and legal permanent residents who were born abroad, not of American parents, and who report they attended high school outside the 50 U.S. states, but earned their highest or most recent degree in the U.S. and (for NSCG) either: (a) entered as a temporary resident on a student/training visa, or (b) entered on a permanent resident, dependent, or other non-work visa but indicated "education" was one of the reasons they first came to the U.S. for a period of six months or more. Excludes 52 respondents who recently earned a U.S. degree, but who both originally and at time of survey were on work-related visas, and who answered negatively to the question whether education was one of their reasons for coming.
Foreign-Born: Immigrated as Adult, All Others	Naturalized U.S. citizens and legal permanent residents who were born abroad, not of American parents, and whose highest and most recent degrees were earned outside the U.S., unless (for NSCG) they originally entered the U.S. on a temporary work visa or as a permanent resident age 21 or older, having earned at least their first bachelor's degree from a foreign institution, and saying that education was not one of their reasons for coming.
Foreign-Born: Temporary Residents, E-2 Treaty Countries	Non-U.S. citizens "with a Temporary U.S. Resident Visa" from certain countries with which the U.S. maintains a free trade or bilateral investment treaty (see State Department website for list of Treaty Countries), in effect prior to 2008. For foreign temporary residents with missing values for the country of citizenship variable, we proxy with country of birth.
Foreign-Born: Temporary Residents, All Others	Non-U.S. citizens "with a Temporary U.S. Resident Visa" from countries with no E-2 designation on the State Department Treaty list, including citizens of some countries that participate only in the E-1 Treaty Trader program.

- Attended high school in the U.S. or abroad
- Earned bachelor's degree from a U.S. or foreign institution
- Earned highest or most recent degree in the U.S., in what year and in what field, as recent STEM graduates may apply to extend their postgraduate work authorization in the U.S.
- Current full- or part-time enrollment in a U.S. degree-granting program
- If married, whether their spouse is working full-time
- Principal employer is U.S. versus foreign
- Employer sector (e.g., academia, industry, nonprofit)
- Occupation (e.g., postsecondary teacher, non-clinical medical scientist)

Based on correlations between these responses and NSCG respondents' first visa type, we identified 10 distinct groups of participants in the U.S workforce. Upon further analysis, however, we found little difference (especially in fully specified models) for U.S. citizens born and raised within the 50 U.S. states, versus those raised abroad. In addition, as previously noted, relatively few foreign temporary resident respondents on student/trainee visas or dependent visas participated in entrepreneurship. For foreign temporary residents the E-2 treaty status of their sending country was substantially more predictive of entrepreneurial activity than their visa category, perhaps in part because both the H-1B and the E-series entrepreneur-investor visas might be categorized by a respondent as "work-related," though the latter might also be coded as "other." Given these empirical findings and related survey constraints, we therefore focus most of the following econometric analysis on six citizenship and immigration categories, as described in Table 3.

COUNTRY-OF-ORIGIN CHARACTERISTICS: GLOBAL ENTREPRENEURSHIP MONITOR & UNESCO

The Global Entrepreneurship Monitor (GEM) Adult Population Survey (APS) collects information on entrepreneurship-related activities and attitudes across more than 100 countries, with typically at least 2,000 adults surveyed per participating country each year. Because not all countries collect data in any given year, for cross-national comparisons researchers often average the responses by country across multiple survey years. For example, in constructing their 2013 Global Entrepreneurship Index (GEI), Ács et al. (2013) averaged GEM APS survey responses for 2010 and 2011 for countries for which both years were available. For this report, we incorporate Ács et al.'s (2013) best estimates of two GEI indicators: cultural support and opportunity perception foreign-born workers' countries-of-origin.

Cultural support is the average of two APS responses, measuring the percentage of adults in each country who agree with the following statements:

- In <country>, most people consider starting a new business a desirable career choice; and
- In <country>, those successful at starting a new business have a high level of status and respect.

We anticipate that immigrants who come to the U.S. as adults to start a business may partly be motivated to do so because their home country offers relatively lower cultural support for entrepreneurship. By contrast, immigrants who choose to come to the U.S. *despite* their home country having relatively high levels of cultural support for entrepreneurship should be relatively *less* likely than other immigrants to engage in entrepreneurship themselves. We therefore construct our explanatory variable *Lack of Cultural Support* as the difference between the U.S. value, and the value for the foreign-born worker's country-of-origin. *Lack of Cultural Support* is therefore set to zero for native U.S. citizens, except for those from Puerto Rico, for whom separate GEM and UNESCO estimates were available.

Opportunity perception is defined as the percentage of adults ages 18 to 64 in the country who said they see good opportunities for starting a business in the area where they live. Because our interest is specifically in entrepreneurship among the college-educated workforce, alone this measure is arguably of limited usefulness. We therefore included an indicator of availability of skilled labor both as an additional control and as an interaction term with opportunity perception. For this purpose we used UNESCO's Institute for Statistics country-level estimates of the percentage of the total population within five years (by age) of secondary school completion who are enrolled in tertiary education.

Prior studies such as that by Clark et al. (2007) have indicated shared language is a significant predictor of immigration. Intuitively, native English speakers may perceive relatively lower costs to immigrate to the U.S. However, the effect of shared language on immigrants' rates of entrepreneurship in their receiving country is theoretically ambiguous. On the one hand, native English speakers' greater facility and fluency may make it easier for them to identify resources, attract financing, and so on. At the same time, because these immigrants have fewer barriers to overcome in the first place, many may be relatively less motivated or attracted by other aspects of participating in the U.S. economy, including the nation's cultural support for entrepreneurship. Empirically, it is also very difficult to disentangle the effect of shared language from E-2 treaty country status, as the vast majority of countries with English as an official language are also E-2 treaty countries. However, another recent study demonstrates no significant effect of coming from an English-speaking country after controlling for other individual- and country-specific characteristics (Blume-Kohout 2015b).

Descriptive Statistics

The share of native U.S. citizens holding STEM postsecondary degrees is declining, as is the share of U.S. STEM occupations held by native U.S. citizens. From 1990 to 2000, the foreign-born share of U.S. resident workers in STEM occupations grew from 11% to 17% at the bachelor's degree level, from 19% to 29% at the master's degree level, and from 24% to 38% at the PhD level (Freeman 2006). In this sub-section, we compare rates of employment in STEM occupations and STEM degree attainment among college-educated native U.S. and foreign-born workers. Then, we describe differences in entrepreneurial participation by higher education attainment and across countries-of-origin. Finally, we summarize educational and demographic characteristics for the full sample (representing the entire U.S. college-educated workforce) and compare these with characteristics of all self-employed individuals, all business owners, STEM business owners, and STEM new venture joiners.

STEM Employment of U.S. versus Foreign-Born College Graduates

Over the past decade, an estimated 44% of employment-based legal immigrants in the United States worked in STEM occupations (Wasem 2012). Immigrants tend more often to work in engineering occupations, where they earn higher wages than native-born U.S. citizens due to their stronger and field-specific academic preparation, including quantitative and analytical skills (Peri and Sparber 2011).

Table 4. Descriptive Statistics

	Full Sample, College-Educated Workforce	All Self-Employed &Business Owners	All Business Owners	STEM Business Owners	STEM New Venture Joiners
U.S. Native	81.3%	80.7%	79.8%	72.2%	69.3%
Foreign-Born:					
Immigrated as Child	4.1%	4.2%	4.4%	4.3%	4.7%
Immigrated as Adult/Student, Highest Degree in US	4.9%	5.1%	5.3%	9.9%	7.2%
Immigrated as Adult, Highest Degree outside U.S.	6.9%	8.3%	8.4%	10.0%	10.3%
Temporary Residents	2.8%	1.7%	2.0%	3.1%	6.5%
Highest Educational Attainment:					
Bachelor's	66.0%	75.0%	77.4%	69.4%	64.3%
Master's or Professional	29.0%	21.9%	20.0%	25.3%	25.5%
Doctorate	5.0%	3.1%	2.6%	4.3%	10.2%
STEM Highest Degree	45.9%	42.7%	44.4%	67.7%	67.9%
US-Earned MBA or Similar	5.2%	4.8%	5.2%	6.1%	8.2%
Female	44.7%	37.3%	34.7%	23.0%	26.8%
Married	73.6%	80.0%	80.6%	80.6%	75.8%
Age (average)	44.3 yrs	48.5 yrs	48.0 yrs	46.8 yrs	37.4 yrs
Experience (average)	16.9 yrs	21.8 yrs	21.5 yrs	20.1 yrs	11.1 yrs

Source: 2010 SESTAT data, including 94,173 observations sampled from U.S. residents holding bachelor's and higher postsecondary degrees, who were employed during the survey reference week(s).

In the 2010 SESTAT data, we find nearly one in four (24.6%) U.S. workers in STEM occupations are foreign-born, including 18.5% of STEM workers with only bachelor's degrees, almost one-third (30.9%) of those with master's or professional degrees, and almost half (47.2%) of those holding doctoral degrees. Put another way, about one-third (34.5%) of college-educated native U.S. workers work in STEM occupations, versus half (50.3%) of U.S. resident, foreign-born workers. STEM occupations with the highest shares of foreign-born immigrant and temporary resident workers include: biochemists and biophysicists (47%); postsecondary teachers of engineering (50%); computer hardware engineers (56%); and non-academic mathematicians (62%).

Yet, among foreign-born workers, there is also a stark difference in STEM workforce participation by citizenship or visa status. Those who immigrated to the U.S. as children are only slightly more likely than native U.S. citizens to work in STEM occupations (37.3%), whereas 59.4% of adult immigrants who came to the U.S. to pursue higher education are STEM workers. Among college-educated adult immigrant workers who earned their highest degree abroad, 43.4% hold STEM jobs. Consistent with the dominant H-1B mechanism for foreign temporary residents from non-E-2 treaty countries, over three-quarters of that group (78.2%) are in STEM occupations, compared with only 57.1% of foreign workers from E-2 treaty countries.

Higher Education and STEM Entrepreneurship

Foreign-born workers are significantly more likely than native U.S. workers to hold their highest degree in a STEM field. Among workers with at least a bachelor's degree, only 2 in 5 (41%) U.S. native workers hold their highest degree in a STEM field, whereas about two-thirds (66%) of immigrant workers hold their highest degree in a STEM field, and another 10.6% hold a STEM bachelor's degree along with some higher degree in a non-STEM field.

As shown in Table 4, over three-quarters (77%) of college-educated business owners hold only a bachelor's degree, but among STEM entrepreneurs advanced degrees are somewhat more common. About a quarter of college-educated STEM business owners hold masters or professional degrees, and another 5% hold doctorates. STEM entrepreneurship is also relatively rare among those who earned only a bachelor's degree in a non-STEM field. Interestingly, among U.S. resident workers with advanced degrees, a higher proportion of STEM entrepreneurs (versus those otherwise employed) hold a graduate degree in a different field than their bachelor's degree. Consistent with Roach and Sauermann's (2014) finding that individuals with a preference for conducting basic research were significantly more likely to be joiners than founders, here we find individuals with PhDs are more strongly represented among STEM new venture joiners. Some 10% of STEM new venture joiners hold PhDs, versus 4.3% of STEM business owners.

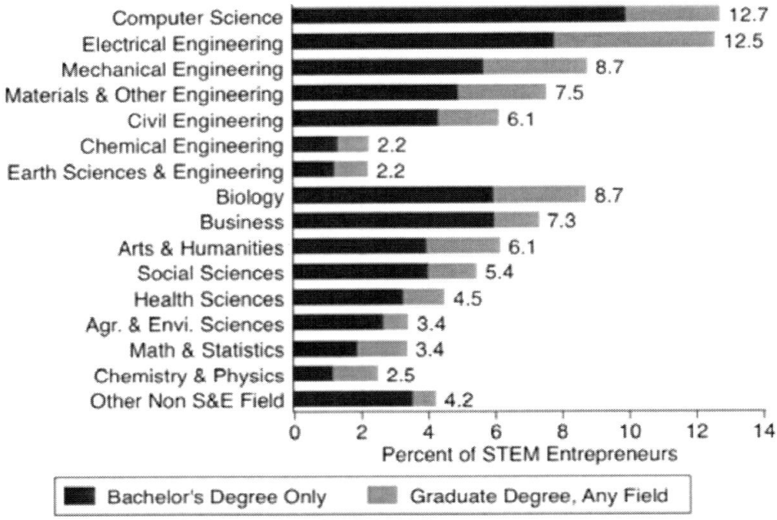

Source: Author's calculations, 2010 SESTAT data.

Figure 3. Bachelor's Degree Fields and Educational Attainment of U.S. STEM Entrepreneurs.

As in Blume-Kohout's (2014) study on gender and STEM entrepreneurship, here we again find individuals with bachelor's degrees in engineering fields generally have higher rates of entrepreneurial participation than those who studied other STEM fields. However, the most common bachelor's degree field among college-educated STEM entrepreneurs overall is computer science. We estimate about 1 in 10 STEM entrepreneurs (9.9%) hold only a bachelor's degree in computer science; another 2.8% of STEM entrepreneurs first earned their bachelor's degree in computer science, but went on to earn a graduate degree in computer science (48%), business (31%), electrical engineering (10%), or some other field. Figure 3 compares the relative frequency of different bachelor's degree fields among STEM entrepreneurs, including both those whose highest educational attainment was that bachelor's degree, as well as those who earned subsequent advanced degrees in any field.

Over half of STEM entrepreneurs who earned their bachelor's degree in chemistry, physics or astronomy also held a graduate degree in some field., and likewise, over 40% of those who majored in chemical engineering, earth sciences and related engineering fields[17], mathematics or statistics. But which advanced degrees did they earn?

Workers who earned their bachelor's degree in physics or astronomy, mathematics or statistics are particularly likely to engage in STEM fields entrepreneurship if they subsequently earned a master's degree in computer science. For workers with bachelor's degrees in these fields, a computer science master's degree may provide additional technical and translational skills that leverage their existing quantitative knowledge and abilities. By contrast, engineering graduate programs tend to require more discipline-specific academic preparation, so may be less accessible as a pathway to entrepreneurship. For those who earned a bachelor's in chemistry, a subsequent degree in electrical engineering is the most common path. On the other hand, individuals who earned their bachelor's degrees in chemical engineering are more likely to engage in entrepreneurship if they also earn an advanced degree in that same field. Those who earned their bachelor's degree in earth sciences or related engineering fields are also more likely to engage in entrepreneurship if they also hold an advanced degree in geology, or in geological and geophysical engineering, however, the range of advanced degrees held by STEM entrepreneurs with a bachelor's degree in this field is broader. Finally, for those who earned their bachelor's degree in biology, an advanced degree in mechanical engineering, mathematics or statistics, or business is positively correlated with entrepreneurship.

STEM Entrepreneurship by Country-of-Origin

Wadwha et al. (2012) estimated that one-third (33.2%) of engineering and technology companies founded by immigrants between 2006 and 2012 had Indian founders, followed by China (8.1%), the United Kingdom (6.3%), Canada (4.2%), Germany (3.9%), Israel (3.5%), Russia (2.4%), Korea (2.2%), Australia (2.0%), and the Netherlands (2.0%).

Although the 2010 SESTAT data indicate only 6.1% of STEM business owners are from India, over 10% of founders (and 28% of foreign-born founders) of high-growth STEM startups started between 2005 and 2010 were from India. Including new venture joiners, we estimate 22.6% of foreign-born STEM entrepreneurs are from India, followed by China (8.1%), Taiwan (5.2%), Canada (4.6%), Korea (4.4%), Vietnam (4.2%), Mexico (4.1%), the Philippines (4.0%), the United Kingdom (2.9%), and Germany (2.0%). Notably, Israel, Australia, and the Netherlands do not appear among the top 10 (or 15) countries-of-origin in our sample.

Figure 4 shows rates of STEM entrepreneurship, broken down by business ownership versus startup venture employment, for the 15 countries that contributed most to U.S. STEM entrepreneurship in 2010. Interestingly, we observe that Filipino workers are substantially less likely to participate in STEM entrepreneurship than native U.S. citizens and all other foreign-born workers from the countries listed, even though they comprise 4.0% of foreign-born STEM entrepreneurs overall. This seeming discrepancy is due to the relatively high share of college-educated immigrant workers in the U.S. overall who came from the Philippines.

Likewise, the substantial contribution of Indian entrepreneurs to U.S. STEM entrepreneurship is partly due to the high share of college-educated immigrant and foreign temporary workers from India. Rates of STEM business ownership, in particular, are notably higher among foreign-born workers from Vietnam, Taiwan, Korea, Russia, Mexico, and Germany, as compared to those from India. Compared to foreign-born workers from these other countries, however, U.S. resident workers from India are much less likely to be naturalized U.S. citizens or legal permanent residents, which limits their avenues for STEM business ownership.

Even so, over 60% of Indian founders of high-growth STEM startups have only temporary resident visas. By contrast, other foreign-born founders of high-growth STEM startups are substantially more likely to have permanent resident visas or be naturalized citizens; less than 10% are foreign temporary residents.

Immigrants from India and China who came to the U.S. as adults to pursue higher education are significantly more likely to engage in STEM entrepreneurship than those who earned their highest degree abroad. Over half (51%) of India-born STEM entrepreneurs earned their highest degree in the U.S. overall, and over 70% of India-born recent founders of high-growth STEM startups earned their degrees in the U.S., as well. For China-born U.S. STEM entrepreneurs, the link with U.S. higher education is even more stark: over three-quarters (77%) of those who founded high-growth STEM startups either earned their highest degree in the U.S. or are currently on student visas, and overall, 65% of China-born STEM entrepreneurs earned their highest degree here.

Hunt (2011) found no difference in probability of having recently founded a startup for foreign-born scientists and engineers with U.S.- versus foreign-earned degrees, and likewise no relationship with age at arrival in the United States. Our results focusing on STEM entrepreneurship strongly differ, and they raise the question: does the U.S. higher education system in some way increase the probability that an individual will become an entrepreneur? Or, is this descriptive result simply driven by the greater proportion of foreign-born workers from China and India who entered on student versus work visas, and the higher propensity towards entrepreneurship among immigrants from those countries?

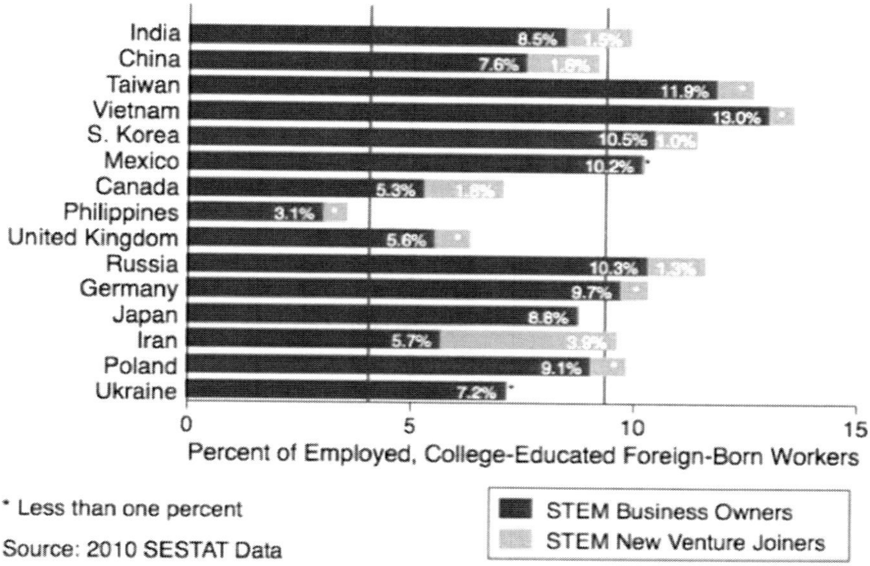

Note: The countries-of-origin are ranked from top to bottom based on each country's total share of STEM business owners. The first vertical line, at about 4.1%, represents the rate of STEM business ownership among native U.S. citizens, excluding those raised in Puerto Rico or U.S. territories. The second vertical line, at about 9.4%, represents the total rate of STEM entrepreneurship (including new venture joiners) for this same group.

Figure 4. Foreign-Born Workers' Rates of Participation in STEM Entrepreneurship, by Country-of-Origin.

ECONOMETRIC MODELS AND RESULTS

In this section, we present results from multivariate logistic regression models predicting entrepreneurial outcomes as a function of demographic characteristics (age, gender, and marital status), educational attainment (highest degree attained, bachelor's degree field, whether highest degree was also in STEM, and so on), postgraduate years of work experience, and country-oforigin characteristics.

Predictors of Business Ownership

In addition to individuals' educational and demographic characteristics, we explore how differences across countries-of-origin may influence college-educated immigrants' rates of entrepreneurial participation. Notionally, we expect adult college-educated immigrants to the United States are more likely to become entrepreneurs when there exists a bigger gap between their home country and the U.S. in terms of cultural support, recognition, and status for entrepreneurs. Because individuals who immigrated as children did not make the same active choice, we might expect such a relationship to be absent or substantially weaker for individuals who immigrated as children. However, we note one indirect pathway remains, due to parental influence. Several prior studies have demonstrated that children of entrepreneurial parents are more likely themselves to become entrepreneurs (Kim, Aldrich, and Keister 2006; Fairlee and Robb 2007). If a parent chose for reasons of greater cultural support to immigrate to the U.S. and pursue entrepreneurial ventures, much as Hout and Rosen (2000) showed for sons of immigrants, we might expect the foreign-born child immigrant also to have greater propensity towards entrepreneurship.

Table 5 presents estimated marginal effects of demographic, educational, immigrant status, and country-of-origin variables on the probability of U.S. business ownership among college-educated U.S. resident workers. These results were obtained from binomial logistic regression on the subsample of employed workers, with the outcome variable equaling 1 if the respondent said their principal employment was with a U.S.-based organization and was best described as either:

- self-employed or a business owner in an incorporated business, professional practice, or farm, or
- same as above, but in an unincorporated business, professional practice, or farm, and either 10 or more employees, or—if fewer than 10 (which could include zero) employees—reporting supervisory responsibility for one or more workers. For all other workers, this outcome was set to zero.

In addition to the variables shown, models also include 17 dummy variables representing different bachelor's degree fields. All variables except bachelor's field of degree are interacted with the citizenship and immigration status categorical variable (see Table 3 for descriptions of these categories), to account for differences across groups in distributions of workers by educational attainment, age, experience, and so on. As shown in Table 5 column (1), we do find significantly higher rates of business ownership, overall, among individuals who immigrated to the U.S. as children, as compared to native U.S. citizens. But, taking country-of-origin characteristics including relative support for entrepreneurship and prevalence of tertiary education into account in model (2), we find that immigrants who arrived in the U.S. as children are significantly less likely to become business owners than their nativity would suggest. Interestingly, adult immigrants who earned their highest degrees outside of the U.S. are also less likely to participate than those who came to the U.S. for higher education, after controlling for country-of-origin characteristics.

U.S. resident workers who hold advanced degrees (including in business administration or management), and those who hold their highest degree in a STEM field are significantly less likely to be business owners than individuals who hold only bachelor's degrees with their highest degree in social sciences or most other non-STEM fields. In particular, those holding only bachelor's degrees in physics or astronomy, mathematics or statistics, or health sciences fields are significantly less likely to become business owners.

Experience—measured here as years since highest degree was completed—is a significant positive predictor of business ownership, with each decade of experience conferring about 3.6 percentage points higher probability of business ownership. Interestingly, for the full sample we observe no additional significant age-related effects. The lack of significance on the age quadratic term for the full sample is partly due to differences in the effects of aging on probability of business ownership across citizenship and immigration status. Note that, because biological age is strongly collinear with career age (experience), especially after highest educational attainment is accounted for, and because we anticipate biological aging has a specific dampening effect on the returns to experience, the econometric estimation includes terms for experience (divided by 10 for convenience in reporting) and biological age squared. However, if we replace this career age "experience" variable with biological age, we can retrieve estimates of predicted probability of business ownership by age that implicitly incorporate experience-related effects. Doing so, we find that for native U.S. citizens and immigrants who arrived in the U.S. as children, the probability of business ownership monotonically increases with age, to at least age 60. Immigrants who arrived as adults similarly show no decline or slowing until at least age 55. By contrast, for foreign temporary residents, business ownership rates begin to decline in the mid-40s, most strongly among those from E-2 treaty countries. However, many of these business owners may have started (or purchased) their companies long ago.

Table 5. Marginal Effects of Demographics, Education, Immigrant Status, and Country-of-Origin Characteristics on Probability of U.S. Business Ownership

Foreign-Born Citizenship/Visa Status:	(1)	(2)
Immigrated as Child	0.0402***	-0.0783**
	(0.0137)	(0.0384)
Immigrated as Adult/Student, Highest Degree in U.S.	0.0571***	0.0340
	(0.0209)	(0.0842)
Immigrated as Adult, Highest Degree outside U.S.	0.0256*	-0.0936***
	(0.0154)	(0.0258)
Temporary Resident, E-2 Treaty Country	0.0195	0.1275
	(0.0386)	(0.1661)
Temporary Resident, All Others	-0.0949***	-0.1627***
	(0.0155)	(0.0373)
Education:		
STEM Highest Degree	-0.0927***	-0.0898***
	(0.0087)	(0.0085)
MBA or Similar U.S. Degree	-0.0372***	-0.0341***
	(0.0126)	(0.0132)
Other Master's or Professional Degree	-0.0792***	-0.0797***
	(0.0061)	(0.0061)
Doctorate	-0.0996***	-0.0992***
	(0.0062)	(0.0063)
Foreign-Born Citizenship/Visa Status:	**(1)**	**(2)**
Demographics:		
Experience/10	0.0382***	0.0359***
	(0.0050)	(0.0050)
Age/10, squared	0.0002	0.0005
	(0.0005)	(0.0005)
Female	-0.0403***	-0.0407***
	(0.0065)	(0.0065)
Married	0.0233***	0.0226***
	(0.0069)	(0.0069)
Country-of-Origin Characteristics:		
Share with Postsecondary Education		-0.0057***
		(0.0008)
Share Perceiving Entrepreneurial Opportunities		0.0013
		(0.0010)
Lack of Cultural Support for Entrepreneurship		0.0289*
		(0.0168)
Number of Observations:	88,631	85,564

Source: 2010 SESTAT data, including 88,631 observations sampled from U.S. resident workers holding bachelor's and higher degrees, and who were employed during the survey reference week(s). Model (2) includes all of the explanatory variables from Model (1), but also incorporates merged country-of-origin characteristics from GEM and UNESCO datasets. All models include dummy variable fixed effects for bachelor's field of degree. Statistical significance: * p< .10, ** p < .05, *** p<.01.

If we focus instead only on owners of startup ventures, the more typical age-related trends emerge, as shown in Figure 5. Still, we note the effects of aging for native U.S. citizens' participation in startup ventures appear much less dramatic. Interestingly, among foreign temporary residents from E-2 treaty countries, there appears to be a slight positive trend, with older individuals more likely recently to have started a business. For most groups, the peak age for starting a business appears to be between 35 and 40.

Consistent with prior literature, we find the rate of business ownership is significantly lower for women. Controlling for education, marital status, nativity and immigration status, we find college-educated women in the U.S. workforce have over 4 percentage points lower probability than men of being business owners.

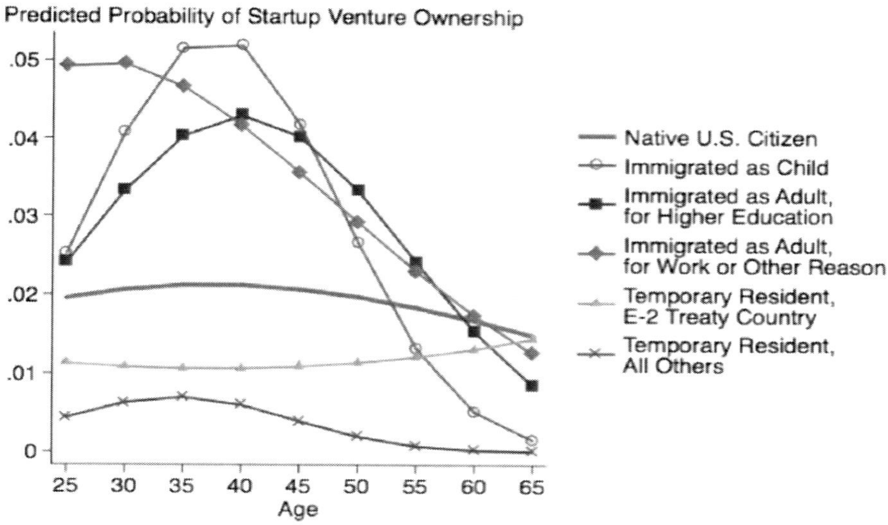

Source: Author's econometric model estimates, 2010 SESTAT data.

Figure 5. Effects of Aging on Startup Ventures, by Citizenship and Immigration Status.

On the other hand, also consistent with prior evidence, we find that marriage is associated with higher probability of business ownership. Though many potential reasons for this have previously been suggested, such as married persons' greater likelihood of having health insurance coverage through their spouse, we note that apparent differences due to marital status may be capturing correlated differences in business ownership by household income and among native U.S. citizens by race/ethnicity. For example, African-Americans in our sample are significantly less likely than non-Hispanic whites to be married, and they are also significantly less likely to be business owners. Separate estimates by citizenship and immigrant status also reveal no significant effect of marriage on entrepreneurship among foreign-born workers who immigrated to the U.S. as adults, nor among foreign temporary residents from non-E-2 treaty countries. For the latter group, restrictions on spousal employment in prevalent visa mechanisms (e.g., H-1B) may preclude advantages like spousal income support or provision of health insurance.

Finally, holding other country-of-origin characteristics constant, we find that foreign-born workers from countries with higher rates of postsecondary education are significantly less likely to become U.S. business owners. This may reflect that when a highly skilled workforce is more broadly available in their home country, there is less incentive for prospective high-skilled entrepreneurs to emigrate. On the other hand, as we hypothesized, immigrants from countries where there is relatively *lower* cultural support for entrepreneurship than in the U.S. are significantly *more* likely to be business owners than otherwise similar workers from countries with relatively higher cultural support.

Demographic, educational, and country-of-origin distributions differ across citizenship/visa groups, as do the effects of some variables that predict entrepreneurship. We control for these differences empirically by interacting all variables except bachelor's field of degree with binary variables for each citizenship/visa category. Although the pooled average marginal effects reported in Table 5 suggest legal permanent residence and citizenship increase probability of business ownership, we can better estimate the effect of a hypothetical change in foreign temporary residents' status—for example, if a new pathway to permanent residence were to become available—by predicting business ownership among foreign temporary residents using their own values of explanatory variables with coefficients estimated for legal permanent residents. In econometrics, this technique of estimating outcomes for one group "treated as" another has a long history, beginning with Oaxaca-Blinder decomposition.

Employing this technique, we estimate that the probability of business ownership for temporary foreign workers from non-E-2 treaty countries would increase by 10 percentage points, from 0.052 to between 0.153 and 0.155 depending on where they earned their highest degree. This difference is highly significant ($p<.0001$ for comparisons with both adult immigrant groups). Although in principle this 10 percentage point estimate could be confounded by differences in English language ability, Blume-Kohout (2015b) shows no effect on business ownership rate of foreign-born workers' coming from countries where English is an official language. This result suggests that alternative visa mechanisms allowing prospective foreign entrepreneurs from non-E-2 countries greater flexibility to start businesses and awarding legal permanent residence for successes might substantially increase participation by the foreign-born in U.S. business ownership.

Predictors of STEM Fields Entrepreneurship

Next, we examined predictors of STEM entrepreneurship in the college-educated U.S.-resident workforce. In addition to restricting the set of business owners described above to those working in STEM occupations, we also include "joiners" in this outcome measure—individuals working in STEM occupations who are employed by startups founded within the previous five years, with fewer than 100 employees.[18] Less than 12% of the STEM entrepreneurs thus defined in our sample are joiners; the rest are business owners. Because the 3.32 percentage point gap in STEM entrepreneurship between foreign-born and native U.S. citizen workers is both statistically and practically more significant than the gap for business ownership overall, in this section we examine the relative contributions of various factors toward this STEM entrepreneurship gap.

Table 6 presents results from a series of five logistic regression models predicting STEM entrepreneurship versus regular employment, with dependent variable equal 1 if the individual is a STEM business owner or joiner, and equal 0 if the person's principal job is with an established U.S.-based firm, non-profit, government agency, university or college, or with a new venture but in a non-STEM occupation or field. The analyses intentionally exclude *non*-STEM business owners from the base reference group.

First, even more strongly than for business ownership overall, we find that different categories of foreign-born workers have different probabilities of STEM entrepreneurship. In particular, as the descriptive statistics depicted in

Figure 2 suggest, there is no statistically significant difference in rates of STEM entrepreneurship for native U.S. citizens versus foreign-born who immigrated to the U.S. as children. In addition, though the sign for foreign temporary residents from non-E-2 countries is positive, the difference is not statistically significant at the 10% level. However, all other groups of foreign-born workers have higher rates of STEM entrepreneurship, most noticeably those who earned their highest degrees from U.S. institutions.

Controlling for differences across these citizenship and immigrant groups in their bachelor's degree fields, as in Table 6 column (2), reduces the apparent gap between native and foreign-born workers by about 60%. For foreign temporary residents from E-2 treaty countries, the probability of STEM entrepreneurship versus regular employment is not significantly different from that for native U.S. citizens. But for foreign temporary residents from *non*-E-2 countries, the sign reverses: controlling for bachelor's field of degree, these foreign temporary residents are significantly *less* likely than native U.S. citizens to be STEM entrepreneurs.

Table 6. Effects of Immigrant Status, Demographics, Education, and Country-of- Origin Characteristics on Probability of U.S. STEM Entrepreneurship

Citizenship/Visa Status:	(1)	(2)	(3)	(4)	(5)
Immigrated as Child	0.0049	0.0003	0.0006	0.0041	0.0035
	(0.0100)	(0.0096)	(0.0095)	(0.0091)	(0.0238)
Immigrated as Adult/Student,	0.0574***	0.0229***	0.0318**	0.0345**	0.0311
Highest Degree in U.S.	(0.0108)	(0.0085)	(0.0134)	(0.0147)	(0.0294)
Immigrated as Adult,	0.0418***	0.0194***	0.0155**	0.0102	-0.0103
Highest Degree outside U.S.	(0.0078)	(0.0062)	(0.0069)	(0.0076)	(0.0160)
Temporary Resident,	0.0425*	0.0200	0.0247	0.0169	0.0410
E-2 Treaty Country	(0.0223)	(0.0175)	(0.0215)	(0.0172)	(0.0482)
Temporary Resident,	0.0080	-0.0197***	-0.0148	-0.0115	0.0095
All Others	(0.0098)	(0.0062)	(0.0104)	(0.0211)	(0.123)
Controls:					
Bachelor's Field of Degree		X	X	X	X
Graduate Degrees			X	X	X
Age, Experience, Gender, Marital Status				X	X
Country-of-Origin Characteristics					X
AIC	103.85	96.27	95.63	93.99	93.79
Number of observations	**78,691**	**78,691**	**78,691**	**78,691**	**78,691**

Marginal effects for probability of STEM business ownership or employment with a new STEM venture versus all other regular employment (excluding non-STEM business owners), estimated using survey-weighted generalized linear models with log link, binomial family. Data are from the 2010 NSF SESTAT database, sampled from U.S. resident workers holding bachelor's and higher postsecondary degrees, and who were employed during the survey reference week(s). Models (1) through (5) incrementally add sets of control variables, as noted in the Table. For Model 5, observations were matched on respondent's country-of-origin to UNESCO and GEM country-level variables. Statistical significance denoted as follows: * p< .10, ** p < .05, *** p<.01.

Table 7. Marginal Effects of Selected Demographic, Educational, and Country-of-Origin Characteristics on Probability of U.S. STEM Entrepreneurship, Adult Immigrants

	Adult Immigrants
Earned Highest Degree in U.S.	0.0262*
	(0.0148)
U.S. STEM Bachelor's Only	0.1157**
	(0.0416)
U.S. Non-STEM Graduate Degree	0.1245***
	(0.0469)
U.S. STEM Master's Degree	0.0851***
	(0.0293)
U.S. STEM Doctorate	0.0434
	(0.0293)
Foreign STEM Bachelor's Only	0.0910***
	(0.0219)
Foreign Non-STEM Graduate Degree	0.0758**
	(0.0324)
Foreign STEM Master's Degree	0.0887***
	(0.0220)
Foreign STEM Doctorate	0.1671***
	(0.0526)
Experience/10	0.0321***
	(0.0110)
Age/10	-0.0187*
	(0.0111)
Female	-0.0428***
	(0.0131)
E-2 Treaty Country	-0.0255**
	(0.0130)
U.S. Degree * Country-of-Origin	0.0012**
Postsecondary Education Share	(0.0005)
Number of Observations	**11,689**

Marginal effects for predicted probability of STEM business ownership or employment with a new STEM venture versus regular employment in an established organization or with a non-STEM venture, estimated using survey-weighted generalized linear models with log link, binomial family. Data are individual-level observations from the 2010 NSF SESTAT database, sampled from U.S. resident workers holding bachelor's and higher postsecondary degrees, and who were employed during the survey reference week(s). Model includes 17 dummy variables representing bachelor's field-ofdegree, and an indicator for marital status, not shown. Delta-method standard errors for the estimated marginal effects are presented in parentheses below the corresponding estimates. Statistical significance denoted as follows: * $p< .10$, ** $p < .05$, *** $p<.01$.

In Table 6 column (3), we see differences in graduate degree attainment explain another 9.3% of the gap for immigrants whose highest degrees were earned abroad. But for immigrants who earned their highest degrees in the U.S., the gap actually widens with addition of these covariates, suggesting U.S. higher education itself may contribute in some way to these immigrants' subsequent entrepreneurship participation. Taken together, differences in highest educational attainment and fields of degree explain approximately half (45%) of the gap in rates of STEM entrepreneurship among U.S.-educated immigrants versus native U.S. citizens. Despite the opposing signs for foreign temporary residents from E-2 versus non-E-2 countries, once higher education degrees are accounted for, neither group has significantly different STEM entrepreneurship participation versus native U.S. citizens.

Controlling for age, experience, gender and marital status, in Table 6 column (4) we find the only group of foreign-born workers that remain with a significantly different rate of STEM entrepreneurship as compared to native U.S. citizens comprises those adult immigrants who earned their highest degrees from U.S. institutions. The rate of STEM entrepreneurship among U.S.-educated adult immigrants is significantly higher at the 10% level than that for foreign-born who arrived in the U.S. as children, and compared to foreign temporary residents from non-E-2 countries. Finally, in Table 6 column (5), we find addition of variables for country-of-origin characteristics including share of population with postsecondary education, relative lack of cultural support for entrepreneurs and its square, and shared English language yield little change in the point estimates. However, collinearity between these variables and the citizenship/immigration status indicators also yields substantially larger standard errors, such that we can no longer detect significant differences in STEM entrepreneurship once these variables are included.

Characteristics of Adult Immigrant Entrepreneurs

In Table 7, we present results from a model similar to that summarized in column (5) of Table 6, but restricting the sample to adult immigrants—that is, naturalized U.S. citizens and legal permanent residents who earned their highest degree in the U.S. or abroad. In this model, we predict probability of STEM entrepreneurship versus regular employment as a function of bachelor's field of degree (17 dummy variables as before, not shown), as well as a more detailed categorical variable for highest degree attained that captures whether the respondent's highest degree was earned in a STEM or non-STEM field. This categorical variable for educational attainment is then interacted with an indicator for having earned that degree in the U.S. versus abroad. As before, demographic variables include age and age-squared, years of experience, female gender, and marital status. Also included are country-of-origin characteristics: share of population in home country with postsecondary education alone and interacted with the indicator for U.S.-earned degree; relative lack of cultural support for entrepreneurship and its square; and an indicator for citizenship with an E-2 treaty country, to test whether the higher rate of entrepreneurship observed among foreign temporary residents from these countries persists once they have immigrated. Because country-of-origin entrepreneurial opportunity perception has been shown insignificant as a predictor of U.S. entrepreneurship, we exclude that variable from this final model.

Overall, earning a degree from a U.S. institution is associated with a 2.6 percentage point increase in the probability of adult immigrants' participating in STEM entrepreneurship. Unsurprisingly, holding a bachelor's degree in some STEM field (and no higher degrees) is also strong positive predictor of STEM entrepreneurship, but this effect is noticeably stronger for U.S.-earned STEM bachelor's degrees, by 2.5 percentage points ($p<.10$). By contrast, immigrants who hold master's or professional degrees in STEM fields participate in STEM entrepreneurship at similar rates, regardless of whether their degree was earned in the U.S. or abroad.

Interestingly, whereas for native U.S. citizens (results not shown) advanced degrees other than the MBA are associated with significantly lower probability of STEM entrepreneurship, for adult immigrants there is no significant negative effect of graduate degrees. In addition, for immigrants who earned their highest degree from a U.S. institution, whether they hold a master's or professional degree in STEM versus non-STEM field appears to be irrelevant. Moreover, among adult immigrants who earned their highest degree abroad, holding a doctorate in a STEM field is associated with a significantly higher rate of STEM entrepreneurship ($p<.10$).

As we observed for business ownership overall, consistent with Blume-Kohout (2014) we find female immigrants are significantly less likely than males to participate in STEM entrepreneurship. Likewise, similar to our overall finding that business ownership increases with experience (Table 5), here we find probability of engaging in STEM entrepreneurship also increases with experience, again defined as years since highest degree awarded. However, aging has a net negative effect, such that rates of STEM entrepreneurship peak around age 50, then decline. Marital status (not shown) has no significant effect. Finally, the significant negative sign on E-2 treaty country-of-origin for naturalized citizens and legal permanent residents—a group whose ability to work and own businesses in the U.S. is not directly affected by the availability of E-2 temporary resident visas—suggests the substantially higher rate of overall business ownership found for foreign temporary residents from these countries is not simply attributable to other latent characteristics, such as shared English language. Interestingly, estimating this same detailed model presented in Table 7 for the subpopulation of foreign temporary residents yields very little of significance, except for approximately 4 percentage points higher probability of STEM entrepreneurship among foreign temporary residents from E-2 treaty countries ($p<.10$).

For the full sample regression in Table 6, we found similar significant effects of countryof-origin postsecondary education and relative lack of cultural support for entrepreneurship as for business ownership overall. However, the subpopulation regression presented in Table 7 finds further correlations between STEM entrepreneurship, rates of postsecondary education in one's home country, and immigrants' decision to pursue U.S. higher education. Here we find that immigrants who came to the U.S. for higher education from countries with relatively high rates of postsecondary education are *more* likely to engage in STEM entrepreneurship, all else equal. Put another way, immigrants who choose to come to the U.S. for higher education *despite* the relatively high prevalence of postsecondary education in their home country may in fact be coming to study in the U.S. disproportionately for economic and entrepreneurial rather than purely academic reasons, perhaps to take advantage of the higher probability of attaining legal permanent residence that a U.S. degree affords. Alternatively, if U.S. higher education itself somehow promotes greater entrepreneurial activity than similar degree programs abroad, prospective immigrant entrepreneurs may implicitly be recognizing this in their decision to earn U.S. degrees.

The net marginal effect of a unit increase in relative lack of cultural support was not statistically significant, reflecting underlying nonlinearity (which we accommodated with a quadratic term in the models). However, logistic regression coefficients were statistically significant for both *Lack of Cultural Support* and its square, with the expected signs. As discussed previously, our expectation is that entrepreneurial individuals will be more likely to emigrate from countries with low cultural support for entrepreneurship, and more likely to immigrate to countries with relatively higher cultural support for entrepreneurship than they found at home. Conversely, if an individual chooses to emigrate from a country with very high cultural support for entrepreneurship, and immigrates to a country with relatively lower levels of support, all else equal we would expect that particular individual to be less inclined towards entrepreneurship.

Figure 6 depicts predicted probabilities of U.S. STEM entrepreneurship by relative cultural support, controlling for differences in sending country's share of population with postsecondary education. As Figure 6 shows, foreign-born workers who immigrated to the U.S. as adults from countries with relatively lower cultural support for entrepreneurship are more likely to engage in STEM entrepreneurship, with more pronounced effects among immigrants who came to the U.S. to pursue higher education versus those who earned their highest degrees abroad. By contrast, among those who immigrated to the U.S. as children, we observe no significant effect of their country-of-origin's characteristics.

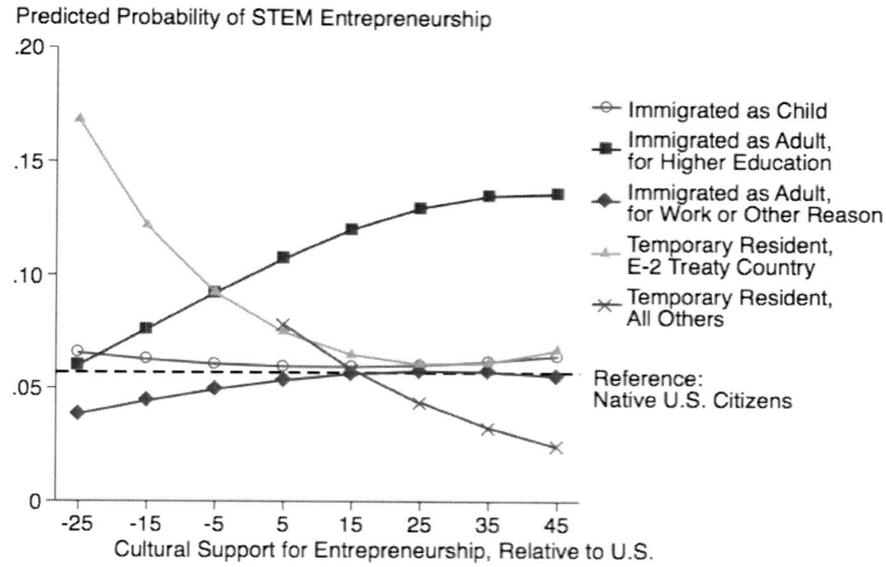

Source: Author's econometric model estimates, 2010 SESTAT and GEM APS data.

Figure 6. Probability of STEM Entrepreneurship by Differences in Country-of-Origin Cultural Support.

For foreign temporary residents the results are inconclusive due to the relatively small number of cases observed, and the trends seem in opposition to our initial intuition. Probing more deeply reveals that the downward-sloping curve for foreign temporary residents from E-2 treaty countries is driven by relatively high rates of U.S. STEM entrepreneurship among Canadian temporary workers. Although Canadian-born workers represent less than 5% of foreign-born U.S. STEM entrepreneurs overall, about 1 in 5 (19%) foreign temporary resident STEM business owners are Canadian. Among all foreign temporary residents who earned their degrees abroad, Canadian foreign temporary residents are more likely than any other group to be STEM business owners instead of organizationally employed. Among foreign temporary residents from E-2 treaty countries, Japanese and Mexicans follow, comprising 11% and 10% of foreign temporary resident STEM business owners, respectively. By contrast, over a third of foreign temporary resident STEM business owners (37%) are Indian, only 6.0% of foreign temporary resident STEM business owners are Chinese, and there are too few other STEM entrepreneurs from non-E-2 countries to extend that curve's trace.

Among adult immigrants to the U.S., the distribution across countries-of-origin in entrepreneurial participation is much broader. Fewer than 1 in 4 (24%) adult immigrant STEM entrepreneurs are from India, with 8.9% and 6.0% from mainland China and Taiwan respectively. Korea and the Philippines follow, each contributing about 4.5%, then the United Kingdom and Russia at about 3.5%.

Finally, less than 15% of STEM entrepreneurs from the United Kingdom and the Philippines, and less than a third of those from Mexico (23%) and Canada (30%), earned their highest degree in the U.S., compared to over two-thirds of STEM entrepreneurs from Korea (74%), China (65%) and Taiwan (92%).

Results from Alternative Estimation Strategies and Measures of Entrepreneurship

To test sensitivity of these results to other more traditional measures of entrepreneurship participation, we re-estimated the models presented in Table 6 with an alternative outcome measure that included all self-employed individuals in STEM—that is, including both STEM business owners and unincorporated non-employer self-employed individuals working in STEM occupations, but excluding STEM new venture joiners. The results were remarkably similar. For example, when we re-estimate the model presented in Table 6 column (4) using this alternative STEM self-employment outcome measure, we find 3.95 percentage points (p<.01) higher probability of STEM self-employment among immigrants who came to the U.S. for higher education, as compared to native U.S. citizens. This result is very similar to the 3.45 percentage point difference we originally estimated. We also again find no significant differences between foreign-born workers who immigrated as children, those who immigrated as adults but earned their highest degree abroad, and foreign temporary residents from E-2 Treaty countries, as compared to native U.S. citizens sharing similar educational and demographic characteristics.

In an alternative model employing multinomial logistic regression estimation, we investigated whether predictors of either joining a startup as an employee or engaging in non-employer unincorporated self-employment differed versus predictors for business ownership, overall. As the descriptive statistics indicate, joiners do tend to be younger than business owners (p<.01), and they are also less likely to be married (p=.013). However, no particular bachelor's field of degree was associated with either higher or lower probability of joining a startup venture as an employee, versus owning a business.

The multinomial logistic regression model also showed that significant positive predictors of choosing regular employment in an established organization over business ownership included being a foreign temporary resident from a non-E-2 Treaty country, working in a STEM occupation, having earned one's highest degree in a STEM field, holding a graduate degree, having fewer years of experience, and being unmarried. We also find in the multinomial model that immigrants who earned their highest degree abroad were significantly more likely (1.25 percentage points, p<.06) than native U.S. citizens to choose employment with a startup venture over business ownership, especially if their highest degree was in a STEM field.

Immigrants who earned a doctorate in any field from a U.S. institution were also more likely to join startup ventures as an employee than as an owner.

To further investigate the possible role of U.S. higher education in spurring foreign-born workers' entrepreneurship, we re-estimated the logistic regression model predicting STEM entrepreneurship as a function of bachelor's field of degree, highest educational attainment, citizenship and immigration status, age, experience, and gender, but now adding 23 dummy variables for country or region of origin,[19] each interacted with a binary indicator for whether the individual earned his or her highest degree from a U.S. institution.

We find that, across foreign-born U.S. workers from most countries and regions, there is a positive association between coming to the U.S. for higher education, versus earning one's highest degree abroad. Foreign-born workers from Israel and most European E-2 Treaty countries, in particular, had a significantly higher probability of STEM entrepreneurship, overall, if they earned their degree from a U.S. institution. The only country-of-origin groups that produced significantly more STEM entrepreneurs overall among individuals who earned their highest degrees abroad instead of from a U.S. institution were native U.S. citizens, and immigrants from the United Kingdom and Mexico.

Finally, like Hunt (2011), in an alternative regression examining probability of having founded a high-growth enterprise (HGE) within the past five years, we observe that foreign-born college-educated workers are significantly more likely, overall, to be HGE founders, as compared to similarly educated native U.S. citizens. Bachelor's field of degree appears to explain none of this relative advantage, but the dampening effect of graduate degrees on the rate of HGEs among native U.S. citizens also does not hold for foreign-born workers. Put another way, among foreign workers who hold at least a bachelor's degree, highest educational attainment seems to be unimportant as a predictor of founding an HGE.

Are Foreign-Born STEM Workers Underemployed?

Kerr (2013) suggests that, among U.S.- and foreign-born STEM workers with similar education, foreign temporary residents are more likely to be underemployed. Given that visa constraints provide foreign temporary residents with less flexibility to seek more appropriate employment in case of a mismatch, we hypothesize that job mismatch may occur more often for these workers, such that they more often find themselves in technician-type jobs that do not provide as much opportunity for expressing their inventive potential. At the very least, foreign temporary residents are likely to be underpaid, relative to similar workers with employer-sponsored permanent residence (i.e., "green cards") (Mukhopadhyay and Oxborrow 2012). In similar vein, inflexibility in employment may preclude not only employee job changes, but also transitions to self-employment and entrepreneurship. In this section, we examine variables relating to job choice, occupational activities, employment, and job satisfaction among foreign-born scientists and entrepreneurs.

In the 2010 SESTAT data, we observe that foreign temporary workers in STEM occupations are significantly more likely than native U.S. citizens to express dissatisfaction with their salary when regularly employed by established organizations. Similar to Hunt's (2015) finding that immigrants with engineering degrees tend to have lower average and 95th percentile wages, we observe substantially lower wages for foreign temporary residents in STEM occupations, especially software developers and programmers. On the other hand, participating in STEM entrepreneurship was associated with significantly higher satisfaction with salaries among foreign temporary residents in STEM occupations. Among all foreign temporary resident workers, only 13.5% of STEM entrepreneurs report dissatisfaction with their salaries, compared with about 18.7% of workers in this group, overall. Given Lan's (2012) related finding that an exogenous increase in permanent resident visas reduced employment of foreign-born, U.S.- trained scientists and engineers' in low-paying postdoc positions, it seems plausible that a mechanism conferring permanent residence visas on this group would potentially decrease the number working for regular, established organizations, and increase movement towards entrepreneurship.

Foreign temporary residents are also significantly more likely than native U.S. citizens to be dissatisfied with their job's level of responsibility. [20] But among foreign temporary residents working in STEM occupations, STEM business ownership (and STEM entrepreneurship more broadly) does not appear to convey the expected positive impact on this measure of satisfaction. In fact, foreign temporary resident STEM business owners from non-E-2 treaty countries are significantly *more* likely to be dissatisfied with their job's level of responsibility, as compared to their regularly employed counterparts. Examining these dissatisfied individuals more closely, we find that over half of them are working as computer systems analysts, systems or network administrators, or software engineers, most of them (66%) holding master's degrees. Many also report they have no management responsibility for people or projects. Interestingly, foreign temporary resident STEM entrepreneurs whose work activities did include managing people or projects were almost three times more likely to be satisfied with their level of responsibility (27.4% versus 9.8%, p=.0524). These findings raise the question, to what extent is the higher rate of self-employment we observe among foreign-born workers driven by necessity or lack of other options, as opposed to entrepreneurial innovation and opportunity recognition?

As of October 2010, we estimate that 3.8% of the college-educated U.S. workforce was unemployed and actively seeking work. Another 3.7% were involuntarily working part-time. For native U.S. citizens, these unemployment and underemployment rates summed to less than 7%, but for immigrants who earned their highest degree abroad, this combined rate exceeded 12%. In addition, about 6% of college-educated native U.S. citizens, versus over 9% of immigrants who earned their highest degrees abroad, also reported they were working in an occupation unrelated to their field of degree because no related job was available.

Immigrants who came to the U.S. as adults are significantly more likely than all other groups of college-educated U.S. workers to indicate dissatisfaction with their job's intellectual challenge, and this distinction by nativity also holds true when we restrict our comparison just to U.S. resident workers employed in STEM occupations in established organizations. Among these regular-employed STEM workers, immigrants are also significantly more likely than native U.S. citizens to report they are dissatisfied with their job's level of responsibility. Conversely, immigrant STEM entrepreneurs reported higher satisfaction with their job's level of responsibility than their regularly employed counterparts. To the extent that foreign-born workers' observable characteristics—for example, current employment, work experience, and educational pedigree—are a poor signal of ability, immigrant workers may have incentive to choose entrepreneurship once their legal status allows (Hegde and Tumlinson 2015). However, dissatisfaction with the conditions of their U.S. employment could also yield declines in highly skilled foreign-born workers' stay rates, regardless of their final immigration status.

Immigrants who earned their highest degrees from U.S. institutions were significantly less likely than other foreign-born workers to be unemployed (4.0%) or underemployed (4.1%), and among those with STEM degrees, less than one-third (32%) reported working in an occupation that did not require STEM technical expertise. Only 4% reported they were working outside their field of degree because no job was available in their field. Nonetheless, about 20% of those in this group working in STEM occupations for established organizations were dissatisfied with their salary, versus 16% among native U.S. citizens and among immigrants who earned their highest degrees abroad. Foreign-born STEM entrepreneurs who earned their highest degree from a U.S. institution also reported greater satisfaction with their job's intellectual challenge and opportunities for advancement, compared to their counterparts working in STEM occupations with established organizations.

Taken together, these results suggest that U.S. higher education degrees may assist adult immigrants in providing a recognizable signal of quality and/or tacit professional acculturation valued by U.S. employers, while potentially also providing better preparation for immigrants' subsequent pursuit of entrepreneurial ventures. On the other hand, discrimination or lack of information among U.S. employers about quality of foreign higher education programs may also contribute to higher rates of unemployment and underemployment we observe among immigrants who earned their degrees abroad (Ferrer and Riddell 2008). Policies that support nascent immigrant STEM entrepreneurs may also help to improve U.S. employment rates, economic productivity, and career satisfaction among new Americans and legal permanent residents. To the extent that U.S. higher education institutions help to prepare workers for entrepreneurial venturing, training and mentoring programs for foreign workers transitioning from H-1B programs who earned their degrees abroad may be a particularly useful target. At present, however, foreign temporary residents' mechanisms and opportunities for STEM entrepreneurship seem unsatisfactory, especially from the perspective of career advancement.

CONCLUSION AND POLICY IMPLICATIONS

At the beginning of this report, we introduced two policy-relevant questions:

1) Why do foreign-born workers have higher rates of participation than native U.S. citizens in entrepreneurship?
2) Why do some foreign-born workers seem to have higher propensity than others towards high-growth, high-impact, science-based entrepreneurship?

U.S.–Foreign STEM Entrepreneurship Gap. As in previous studies, we find a substantial percentage of the gap in rates of business ownership and STEM entrepreneurship among college-educated foreign-born and native U.S. workers can be attributed to differences in higher education attainment, particularly the undergraduate major they choose. Differences in the distribution of U.S. and foreign-born workers by age, experience, gender, and marital status explain most of the remaining gap, except among immigrants who earned their highest degree in the United States. For this group, the rate of STEM entrepreneurship is almost 4 percentage points higher than for otherwise-similar native U.S. citizens.

Foreign-Born + U.S. Educated = Higher STEM Entrepreneurship Rate. Although descriptive statistics suggest contradictory findings for business owners overall, after controlling for other demographic and educational characteristics we find higher rates of STEM entrepreneurship participation among immigrants who earned their highest degrees from U.S. institutions, as compared to both native U.S. citizens and other U.S. resident foreign-born workers. U.S. higher education institutions may serve a role in advancing foreign students' English language skills, certifying quality for prospective employers, and providing professional and entrepreneurial acculturation (Suter and Jandl 2008). Previous studies have demonstrated the importance of graduate and professional training in influencing subsequent entrepreneurship (Bercovitz and Feldman 2008; Blume-Kohout 2014, 2015a; Stuart and Ding 2006). Thus, it seems plausible that U.S. higher education itself may promote greater subsequent entrepreneurial venturing among foreign-born workers. However, the analyses herein cannot rule out the possibility of selection. That is, immigrants who choose to come to the U.S. to pursue higher education degrees may have other unobserved characteristics which predispose them towards entrepreneurship, as compared to immigrants who arrive for work or other reasons.

STEM Entrepreneurship Link to E-2 Treaty Investor Status. As discussed in the Introduction, one limitation of the data used for this study is that it does not include specific information on the respondent's visa type. As such, we cannot definitively identify which visas foreign temporary resident STEM entrepreneurs are covered by. Overall, however, visa constraints for foreign temporary residents—particularly those from countries not eligible for the E-2 Investor visa—seem to decrease dramatically their participation in self-employment in general, and STEM entrepreneurship more specifically. On the other hand, foreign temporary workers from E-2 Treaty countries participate in STEM entrepreneurship at similar rates as native U.S. citizens. Descriptively, workers in this group are also significantly more likely than similar native U.S. citizens to choose business ownership over non-employer self-employment. More research is needed to understand the extent to which each of the various non-immigrant visa types currently available are used by nascent STEM entrepreneurs.

Implications for Policy

At present, the United States remains a destination of choice for achievement-oriented foreign nationals seeking greater cultural support for entrepreneurship than their home countries provide. The results presented in this report suggest several opportunities for policymakers interested in spurring new business creation and startup employment in new ventures, focusing in STEM fields. As of February 2016, the U.S. Department of Homeland Security has issued proposed rules and guidance documents that may update some of the visa categories discussed in this section, and additional proposals are planned.

High-Skilled Immigrant Underemployment. The slowing rate of immigrant participation in STEM entrepreneurship observed in recent years may be attributed to several factors, including possibly lower stay rates among innovative foreign-born workers. For example, the higher rates of dissatisfaction we observed among highly skilled immigrant workers trained abroad with the intellectual challenge and level of responsibility their jobs provide, along with their substantially higher rate of unemployment, suggests room for improvement in matching immigrant STEM workers in the U.S. to jobs that make full use of their skills and abilities. More research is also needed to understand how tacit knowledge, skills, innate ability and aspirations may differ across individuals in this group, and whether dissatisfied individuals may be trading off job satisfaction for social or family-related reasons. Specifically, previous studies have shown the importance of ethnic enclaves in attracting immigrants to communities with shared national and cultural origins, and as a result, some immigrants may preferentially locate in cities that provide that sense of community, even if the job market for someone with their knowledge, skills, abilities and interests is relatively weak.

Post-Graduation Pathways to STEM Entrepreneurship. Given the higher rate of STEM entrepreneurship we observe among immigrants who earned their highest degrees here in the United States, policymakers might be tempted to conclude we should aim to increase foreign student enrollment in STEM fields at U.S. universities. But, prior research suggests policies aimed explicitly at increasing foreign student enrollment in U.S. universities may have unintended negative consequences for enrollment of native U.S. students, at both the undergraduate and graduate levels (Betts and Lofstrom 2008; Hoxby 1998; Borjas 2007). Based on the empirical results contained in this report, it seems a better solution would be to expand pathways for foreign temporary resident students who earn advanced degrees in the U.S. to extend their stay after graduation, contingent on an offer of employment related to their U.S.-earned degree, regardless of employer sector.

The GEIR Program and Support for New Venture Joiners. Initiatives like the Global Entrepreneur-in-Residence (GEIR) program in Massachusetts seem to provide a convenient solution within current legislative and regulatory confines. The GEIR program provides aspiring foreign entrepreneurs access to academic (uncapped) H-1B visas along with mentoring and training. At the same time, we find STEM entrepreneurship strongly increases with years of experience since highest degree. Requiring that graduates with a propensity toward entrepreneurship start their business immediately after graduation or else leave the country may be counterproductive, decreasing the probability of economically productive and satisfying outcomes. Among adult immigrants who earned their highest degree in the U.S., the peak age range for starting a new STEM venture was between 35 and 40. This later peak in participation could partly be due to delays in foreign students' achieving legal permanent residence or U.S. citizenship after graduation, a change in status that would provide them greater flexibility in (self-) employment. Employment in industry and especially with startup ventures may provide important on-the-job training and human capital that encourage foreign-born workers eventually to pursue entrepreneurship themselves.

Expansion of H-1B and F1/OPT Visa Programs.[21] Our results thus support more generous expansion of H-1B visas and/or F-1 Optional Practical Training (OPT) eligibility to foreign students who earned graduate degrees from U.S. institutions and receive offers of employment in related STEM occupations, regardless of the employer's sector or foreign graduates' immediate startup intent. At present, some 20,000 H-1B visas are reserved annually for foreign temporary resident students who earned graduate degrees from U.S. institutions, but only those students graduating with advanced degrees in STEM fields or business are eligible to extend their F-1 student visas via Optional Practical Training (OPT) beyond the initial 12-month period after graduation. Foreign students on temporary resident visas who complete degrees at U.S. institutions in non-STEM fields are not eligible for this F-1 OPT extension. However, among immigrants who came to the U.S. for higher education and earned their highest degrees here, we find that whether their *highest* degree was earned in a STEM field is irrelevant to their probability of engaging in STEM entrepreneurship. Based on this evidence, we suggest that if a foreign student who graduates from a U.S. master's or doctoral program is able to obtain an offer of employment in a H-1B eligible STEM occupation with any U.S. employer (including new startups), s/he should be eligible for the F-1 STEM OPT extension, even if his or her most recent U.S. degree was in a less closely related or non-STEM discipline. In effect, if their prior knowledge, skills, training and/or innate abilities are sufficient to serve in a position requiring bachelor's-level technical expertise in some STEM field (thus consistent with the specialized knowledge requirement of the H-1B), the field of their most recent degree may be irrelevant.

Missouri's Exemplary Entrepreneurship Program. Expanding the F-1 OPT program might also permit further entrepreneurial engagement by recent graduates, as self-employment is permitted. Along these lines, Luppino et al. (2012) note the University of Missouri-Kansas City's Entrepreneurship Scholars program does not restrict participation on the basis of field or prior educational attainment. Combining a similar lack of terminal degree restriction with innovative partnerships like Massachusetts' GEIR program utilizing H-1B visas would thus provide greater opportunities for entrepreneurial graduates to engage in mentored startups, and such mentored engagement might also help to reduce the career dissatisfaction many foreign temporary residents expressed regarding their independent efforts as STEM entrepreneurs.

STEM Workforce Visa. Some advocates of immigration reform have suggested awarding permanent residence conditional on remaining engaged in the STEM workforce, in an occupation related to one's field of degree, for several years after graduation from a U.S. graduate program (Gilgannon 2013; Kauffman Foundation 2015). As Gilgannon (2013) suggests, allowing foreign temporary resident workers with verified bachelor's or higher education earned here or abroad and at least five years' U.S. work experience, including but not limited to those in H-1B status, to remain in the U.S. and apply their on-the-job training and knowledge towards a new venture may be particularly productive, and arguably more so than any mechanism requiring a new graduate immediately to embark on a startup venture, else leave the country.

Startup Visa.[22] Legislative initiatives in recent years have aimed towards a true "startup visa," which would allow foreign-born entrepreneurs to achieve permanent residence upon satisfying specified conditions during an initial conditional residence period similar to the EB-5. One such approach would be to require applicants to demonstrate they have attracted substantial U.S. investment in their business, such as from a U.S.-based venture capital firm, angel investor, or government agency (Gilgannon 2013). Then, by the end of the two-year initial conditional residence period, the venture must demonstrate that it has created new jobs, and also has achieved either high revenue growth, or alternatively that it has attracted subsequent rounds of investment. In contrast with the EB-5, this path would not select on personal wealth, but rather on market assessment of the potential economic value of the product or service to be provided.

E-2 Investor Visa Expansion. The clear advantage of this pathway for attracting STEM-based, high-tech entrepreneurship to the U.S. is in its use of an intermediate benchmark—in this case, venture funding—while research and development, regulatory clearances, and similar steps are still underway. In this current report, we found evidence that the similarly flexible E-2 Treaty Investor visa is associated with significantly higher rates of STEM entrepreneurship among foreign temporary residents from eligible countries, with participation rates comparable to that among native U.S. citizens after controlling for workers' demographic characteristics and education. But one drawback of the current E-2 visa is the requirement that the applicant him- or herself be capable of a substantial up front financial investment, which may be infeasible for younger entrepreneurs or high-tech and R&D-based businesses. Expanding opportunities for permanent residence—or, at least, unlimited renewal of temporary resident status—conditional on benchmark successes, and opening eligibility to citizens of India and other countries not eligible for the E-2 visa, seems likely to increase U.S. STEM entrepreneurship overall.

International Competition for STEM Graduates. As mentioned before, the visa-degree gap may put the U.S at a disadvantage vis-à-vis other advanced countries. Citizens of nations like Canada and Australia that combine a highly educated labor force with strong cultural support for entrepreneurship are already unlikely to immigrate to the U.S. to start entrepreneurial ventures, and increasingly these countries are implementing policies designed to attract and retain highly skilled innovative and entrepreneurial foreign workers. For example, Canada, Australia, and New Zealand all use a points-based scheme for migrant visa eligibility which awards additional points to applicants who have at least two years' study in their countries' higher education institutions (Suter and Jandl 2008). Canada also lowers the threshold for number of points required for a prospective immigrant classified as an entrepreneur, "who intends and has the ability to establish... business or commercial venture in Canada that will make a significant contribution to the economy" (Tabag 2013). Since 2001, Australia also has allowed foreign students to remain and work for any employer during their wait for legal permanent residence. With such a large gap between the number of foreign students graduating from U.S. higher education institutions and the number of employment-based visas available to highly skilled graduates, along with evidence of underemployment and related job dissatisfaction among immigrants who earned their degrees abroad, the U.S. risks driving trained, innovative, and entrepreneurial talent offshore.

References

Ács, Zoltán J., László Szerb, and Erkko Autio. 2013. *Global Entrepreneurship and Development Index 2013*. Northampton, MA: Edward Elgar.

Anderson, Stuart. 2013. *American Made 2.0: How Immigrant Entrepreneurs Continue to Contribute to the U.S. Economy*. Arlington, VA: National Venture Capital Association.

Åstebro, Thomas, Jing Chen, and Peter Thompson. 2011. "Stars and misfits: self-employment and labor market frictions." *Management Science* no. 57 (11):1999-2017.

Belot, Michèle V.K., and Timothy J. Hatton. 2012. "Immigrant selection in the OECD." *Scandinavian Journal of Economics* no. 114 (4):1105-1128.

Bercovitz, Janet, and Maryann Feldman. 2008. "Academic entrepreneurs: organizational change at the individual level." *Organization Science* no. 19 (1):69-89.

Betts, Julian R., and Magnus Lofstrom. 2008. "The educational attainment of immigrants: trends and implications." In *Issues in the Economics of Immigration*, edited by George J. Borjas, 51-115. Chicago, IL: University of Chicago Press.

BLS. 2012. Attachment C: Detailed 2010 SOC occupations included in STEM SOC Policy Committee recommendation to OMB. In *2010 SOC User Guide*. Washington, DC: U.S. Bureau of Labor Statistics.

———. 2015. *Table 7. Survival of private sector establishments by opening year*. U.S. Bureau of Labor Statistics, 19 November [cited 9 February]. Available from http://www.bls.gov/bdm/us_age_naics_00_table7.txt.

Blume-Kohout, Margaret E. 2014. *Understanding the Gender Gap in STEM Fields Entrepreneurship*. SBA Office of Advocacy Report no. 424. Washington, DC: U.S. Small Business Administration. Available at http://www.sba.gov/advocacy/understanding-gender-gap-stem-fields-entrepreneurship.

———. 2015a. Gender Differences in the Influence of Training Environments on Entrepreneurship. Paper read at Society of Labor Economists Annual Meeting and Fourth SOLE/EALE World Conference, at Montreal, Canada.

———. 2015b. Why are some foreign-born workers more entrepreneurial than others? *Journal of Technology Transfer* in press: 1-27.

Blume-Kohout, Margaret E., and Dadhi Adhikari. 2014. Training the Biomedical Workforce: Does Funding Mechanism Matter? Paper read at American Economic Association Annual Meeting, 3-5 January, at Philadelphia, PA.

Borjas, George J. 1987. "Self-selection and the earnings of immigrants." *American Economic Review* no. 77 (4):531-553.

———. 2002. *An Evaluation of the Foreign Student Program*. Washington, DC: Center for Immigration Studies. Available at http://www.cis.org/ForeignStudentProgram.

———. 2007. "Do foreign students crowd out native students from graduate programs?" In *Science and the University*, edited by Paula E. Stephan and Ronald G. Ehrenberg, 134-149. Madison, WI: University of Wisconsin Press.

Bound, John, Breno Braga, Joseph M. Golden, and Gaurav Khanna. 2015. "Recruitment of foreigners in the market for computer scientists in the United States." *Journal of Labor Economics* no. 33 (S1):S187-S223.

Brücker, Herbert, and Cécily Defoort. 2009. "Inequality and the self-selection of international migrants: theory and new evidence." *International Journal of Manpower* no. 30 (7):742-764.

Chellaraj, Gnanaraj, Keith E. Maskus, and Aaditya Mattoo. 2008. "The contribution of international graduate students to U.S. innovation." *Review of International Economics* no. 16 (3):444-462.

Clark, Ximena, Timothy J. Hatton, and Jeffrey G. Williamson. 2007. "Explaining U.S. immigration, 1971-1998." *Review of Economics and Statistics* no. 89 (2):359-373.

Doran, Kirk, Alexander Gelber, and Adam Isen. 2015. The effects of high-skilled immigration policy on firms: evidence from H-1B Visa Lotteries. Paper read at NBER Summer Institute, June 2015, at Cambridge, MA.

Drori, Israel, Benson Honig, and Mike Wright. 2009. "Transnational entrepreneurship: an emergent field of study." *Entrepreneurship Theory and Practice* no. 33 (5):1001-1022.

Dugan, Andrew. *Passing New Immigration Laws Is Important to Americans*. Gallup. Available from http://www.gallup.com/poll/163475/passing-new-immigration-laws-importantamericans.aspx?utmsource=&utm medium=& utm campaign=tiles.

Elfenbein, Daniel W., Barton H. Hamilton, and Todd R. Zenger. 2010. "The small firm effect and the entrepreneurial spawning of scientists and engineers." *Management Science* no. 56 (4):659-681. doi: 10.1287/mnsc.1090.1130.

Fairlee, Robert W. 2008. *Estimating the Contribution of Immigrant Business Owners to the U.S. Economy*. SBA Office of Advocacy Report no. 334. Washington, DC: U.S. Small Business Administration. Available at http://archive.sba.gov/advo/research/rs334tot.pdf.

———. 2012. *Immigrant Entrepreneurs and Small Business Owners, and their Access to Financial Capital*. SBA Office of Advocacy Report no. 396.

Washington, DC: U.S. Small Business Administration. Available at http://www.sba.gov/sites/default/files/rs396tot.pdf.

———. 2014. *Kauffman Index of Entrepreneurial Activity: 1996-2013*. Kansas City, MO: Ewing Marion Kauffman Foundation. Available at http://www.kauffman.org/~/media/kauffman_org/research reports and covers/2014/04/kiea 2014 report.pdf.

Fairlee, Robert W., and Alicia Robb. 2007. "Families, human capital, and small business: evidence from the characteristics of business owners survey." *Industrial & Labor Relations Review* no. 60 (2):225-245.

Ferrer, Ana, and W. Craig Riddell. 2008. "Education, credentials, and immigrant earnings." *Canadian Journal of Economics* no. 41 (1):186-216.

Finn, Michael G. 2010. *Stay Rates of Foreign Doctorate Recipients from U.S. Universities, 2007*. Oak Ridge, TN: Oak Ridge Institute for Science and Education.

Freeman, Richard B. 2006. "Does globalization of the scientific/engineering workforce threaten US economic leadership?" *Innovation Policy and the Economy* no. 6:123-157.

Gilgannon, Stuart D.P. 2013. "The land of opportunity: why more must be done to encourage immigrant entrepreneurship in the United States." *Duquesne Business Law Journal* no. 15 (1):1-26.

Grogger, Jeffrey, and Gordon H. Hanson. 2015. "Attracting talent: location choices of foreign-born PhDs in the United States." *Journal of Labor Economics* no. 33 (S1):S5-S38.

Hart, David M., and Zoltán J. Ács. 2011. "High-tech immigrant entrepreneurship in the United States." *Economic Development Quarterly* no. 25 (2):116-129.

Hegde, Deepak, and Justin Tumlinson. Department. 2015. Unobserved ability and entrepreneurship. 20 April 2015 [Last accessed 14 July 2015]. Available from http://ssrn.com/abstract=2596846.

Hout, Michael, and Harvey S Rosen. 2000. "Self-employment, family background, and race." *Journal of Human Resources* no. 35 (4):670-692.

Hoxby, Caroline M. 1998. "Do immigrants crowd disadvantaged American natives out of higher education?" In *Help or Hindrance? The Economic Implications of Immigration for African Americans*, edited by Daniel S. Hammermesh and Frank O. Bean, 282-381. New York: Russell Sage Foundation.

Hsu, David H., Edward B. Roberts, and Charles E. Eesley. 2007. "Entrepreneurs from technology-based universities: evidence from MIT." *Research Policy* no. 36 (5):768-788.

Hunt, Jennifer. 2011. "Which immigrants are most innovative and entrepreneurial? Distinctions by entry visa." *Journal of Labor Economics* no. 29 (3):417-457.

———. 2015. "Are immigrants the most skilled US computer and engineering workers?" *Journal of Labor Economics* no. 33 (S1):S39-S77.

Hunt, Jennifer, and Marjolaine Gauthier-Loiselle. 2010. "How much does immigration boost innovation?" *American Economic Journal-Macroeconomics* no. 2 (2):31-56.

Kahn, Shulamit, Giulia La Mattina, and Megan J. MacGarvie. 2015. "Misfits," "stars," and immigrant entrepreneurship. Paper read at Society of Labor Economists Annual Meeting, May 2012, at Chicago, IL.

Kahn, Shulamit, and Megan J. MacGarvie. 2014. How important is US location for research in science? *Review of Economics and Statistics* in press.

Kato, Takao, and Chad Sparber. 2013. "Quotas and quality: the effect of H-1B visa restrictions on the pool of prospective undergraduate students from abroad." *Review of Economics and Statistics* no. 95 (1):109-126.

Kauffman Foundation. 2015. *Immigrant Entrepreneurs: A Path to U.S. Economic Growth.* Entrepreneurship Policy Digest Kansas City, MO: Ewing Marion Kauffman Foundation. Available at http://www.kauffman.org/~/media/kauffman_org/resources/2015/entrepreneurshippolicydigest/january2015/entrepreneurship_policy_digest_january_2015_immigrant_entrepreneurs_pathtouseconomicgrowth.pdf.

Kerr, Sari Pekkala, William R. Kerr, and William F. Lincoln. 2015a. "Firms and the economics of skilled immigration." *Innovation Policy and the Economy* no. 15.

———. 2015b. "Skilled immigration and the employment structures of US firms." *Journal of Labor Economics* no. 33 (S1):S147-S186.

Kerr, William R. Department. 2013. *U.S. High-Skilled Immigration, Innovation, and Entrepreneurship.* Harvard Business School Working Paper 14-017. Cambridge, MA. Available from http://www.people.hbs.edu/wkerr/Kerr_HighSkillImmEmpirics-8-26-13.pdf.

Kerr, William R., and William Lincoln. 2010. "The supply side of innovation: H-1B visa reforms and US ethnic invention." *Journal of Labor Economics* no. 28 (3):473-508.

Kim, Phillip H., Howard E. Aldrich, and Lisa A. Keister. 2006. "Access (not) denied: the impact of financial, human, and cultural capital on entrepreneurial entry in the United States." *Small Business Economics* no. 27 (1):5-22.

Lan, Xiaohuan. 2012. "Permanent visas and temporary jobs: evidence from postdoctoral participation of foreign PhDs in the United States." *Journal of Policy Analysis and Management* no. 31 (3):623- 640.

Lazear, Edward P. 2005. "Entrepreneurship." *Journal of Labor Economics* no. 23 (4):649-680.

Luppino, Anthony, John Norton, and Malika Simmons. 2012. *Reforming Immigration Law to Allow More Foreign Student Entrepreneurs to Launch Job-Creating Ventures in the United States*. Kansas City, MO: Ewing Marion Kauffman Foundation. Available at http://www.kauffman.org/~/media/kauffman_org/researchreportsandcovers/2012/08/kauffmanimmigrationreform.pdf.

Matloff, Norman. 2013. *Are Foreign Students the 'Best And Brightest'? Data and Implications for Immigration Policy*. Washington, DC: Economic Policy Institute. Briefing Paper #356. Available at http://www.epi.org/publication/bp356-foreign-students-best-brightest-immigration-policy/.

Mukhopadhyay, Sankar, and David Oxborrow. 2012. "The value of an employment-based green card." *Demography* no. 49 (1):219-237.

No, Yeonji, and John P. Walsh. 2010. "The importance of foreign-born talent for U.S. innovation." *Nature Biotechnology* no. 28:289-291.

Olney, William W. 2013. "Immigration and firm expansion." *Journal of Regional Science* no. 53 (1):142- 157.

Orrenius, Pia M., and Madeline Zavodny. 2015. "Does immigration affect whether US natives major in science and engineering?" *Journal of Labor Economics* no. 33 (S1):S79-S108.

Peri, Giovanni, Kevin Shih, and Chad Sparber. 2015. "STEM workers, H-1B visas, and productivity in US cities." *Journal of Labor Economics* no. 33 (S1):S225-S255.

Peri, Giovanni, and Chad Sparber. 2011. "Highly educated immigrants and native occupational choice." *Industrial Relations* no. 50 (3):385-411. doi: 10.1111/J.1468-232x.2011.00643.X.

Roach, Michael, and Henry Sauermann. 2014. "Founder or joiner? The role of preferences and context in shaping different entrepreneurial interests." *Management Science* no. forthcoming.

Ruiz, Neil G. 2014. *The Geography of Foreign Students in U.S. Higher Education: Origins and Destinations*. Global Cities Initiative Washington DC: Brookings Institution. Available at http://www.brookings.edu/~/media/research/files/reports/2014/08/foreignstudents/foreignstudentsfinal.pdf.

Salzman, Hal, Daniel Kuehn, and B. Lindsay Lowell. 2013. *Guestworkers in the high-skill U.S. labor market.* Briefing Paper #359. Washington, DC: Economic Policy Institute. Available at http://www.epi.org/publication/bp359-guestworkers-high-skill-labor-market-analysis/.

Saxenian, AnnaLee. 1999. *Silicon Valley's New Immigrant Entrepreneurs.* San Francisco, CA: Public Policy Institute of California. Available at http://www.ppic.org/content/pubs/report/R 699ASR.pdf.

Stephan, Paula E., and Sharon G. Levin. 2001. "Exceptional contributions to US science by the foreign-born and foreign-educated." *Population Research and Policy Review* no. 20 (1/2):59-79.

———. 2007. "Foreign scholars in U.S. science: contributions and costs." In *Science and the University*, edited by Paula E. Stephan and Ronald G. Ehrenberg, 113-133. Madison, WI: University of Wisconsin Press.

Stuart, Toby E., and Waverly W. Ding. 2006. "The social structural antecedents of commercial activity in the life sciences." *American Journal of Sociology* no. 112 (1):97-144.

Stuen, Eric T., Ahmed Mushfiq Mobarak, and Keith E. Maskus. 2012. "Skilled immigration and innovation: evidence from enrolment fluctuations in US doctoral programmes." *The Economic Journal* no. 122 (December):1143-1176.

Suter, Brigitte, and Michael Jandl. 2008. "Train and retain: national and regional policies to promote the settlement of foreign graduates in knowledge economies." *International Migration & Integration* no. 9:401-418.

Tabag, Carla. 2013. "The point of a points system: attracting highly skilled immigrants the United States needs and ensuring their success." *Vanderbilt Journal of Transnational Law* no. 46 (1):271-302.

U.S. Department of State. 2014. *Report of the Visa Office 2014.* Washington, DCAvailable at http://travel.state.gov/content/dam/visas/Statistics/Annual Reports/FY2014AnnualReport/FY14An nualReport-TableVI-PartIV.pdf.

UCSIS. *Questions & Answers: USCIS Issues Guidance Memorandum on Establishing the "Employee-Employer Relationship" in H-1B Petitions.* U.S. Citizenship and Immigration Services, U.S. Department of Homeland Security, 12 March 2012 [cited 6 February 2015]. Available from http://www.uscis.gov/news/public-releases-topic/business-immigration/questions-answers-uscisissues-guidance-memorandum-establishing-employee-employer-relationship-h-1b-petitions.

USCIS. 2010. Determining Employer-Employee Relationship for Adjudication of H-1B Petitions, Including Third-Party Site Placements. Washington, DC:

U.S. Citizenship and Immigration Services, Department of Homeland Security.

Wadwha, Vivek, Ben Rissing, AnnaLee Saxenian, and Gary Gereffi. 2007. *Education, Entrepreneurship and Immigration: America's New Immigrant Entrepreneurs, Part II.* Kansas City, MO: Ewing Marion Kauffman Foundation. Available at http://www.kauffman.org/~/media/kauffman_ org/research reports and covers/2007/06/entrep_immigrants_2_61207.pdf.

Wadwha, Vivek, AnnaLee Saxenian, Richard Freeman, and Gary Gereffi. 2007. *America's Loss is the World's Gain: America's New Immigrant Entrepreneurs.* Kansas City, MO: Ewing Marion Kauffman Foundation. Available at http://www.kauffman.org/~/media/kauffman_org/research reports and covers/2009/03/americas_loss.pdf.

Wadwha, Vivek, AnnaLee Saxenian, and F. Daniel Siciliano. 2012. *Then and Now: America's New Immigrant Entrepreneurs, Part VII.* Kansas City, MO: Ewing Marion Kauffman Foundation. Available at http://www.kauffman.org/~/media/kauffman_org/research reports and covers/2012/10/then and now americas new immigrant entrepreneurs.pdf.

Wasem, Ruth Ellen. 2012. *Immigration of Foreign Nationals with Science, Technology, Engineering, and Mathematics (STEM) Degrees.* Washington, DC: Congressional Research Service. Available at http://fas.org/sgp/crs/misc/R42530.pdf.

Webber, Karen L. 2012. "Research productivity of foreign- and US-born faculty: differences by time on task." *Higher Education* no. 64 (5):709-729.

Winston, Gordon C. 1999. "Subsidies, hierarchy and peers: the awkward economics of higher education." *Journal of Economic Perspectives* no. 13 (1):13-36.

End Notes

[1] In this report, the terms "foreign-born" and "immigrant" refer to different groups of U.S. residents. "Foreign-born" includes all U.S. residents who are not native U.S. citizens. "Immigrant" refers to the subset of foreign-born U.S. residents who have attained either naturalized U.S. citizenship or legal permanent residence, and thus it excludes non-immigrant foreign temporary residents on student, work, or dependent visas.

[2] The E-2 nonimmigrant classification allows a national of a treaty country (a country with which the United States maintains a treaty of commerce and navigation) to be admitted to the United States when investing a substantial amount of capital in a U.S. business. Certain employees of such a person or of a qualifying organization may also be eligible for this classification. U.S. Citizenship and Immigration Service website, www.uscis.gov/working-united-states/temporary-workers/e-2-treaty-investors, accessed January 29, 2016.

[3] For a comprehensive discussion, see the U.S. Citizenship and Immigration Service's "Entrepreneur Visa Guide," available at: www.uscis.gov/eir/visa-guide/entrepreneur-visa-guide, last accessed July 24, 2015.

[4] U.S. Citizenship and Immigration Services, "Understanding F-1 OPT Requirements," at http://www.uscis.gov/eir/visa-guide/f-1-opt-optional-practical-training/understanding-f-1-opt-requirements, last accessed 6 February 2015. See also U.S. Department of Homeland Security, "Working in the United States," at http://studyinthestates.dhs.gov/working-in-the-united-states, and "International Students and Entrepreneurship," at http://studyinthestates.dhs.gov/international-students-and-entrepreneurship, both last accessed July 24, 2015.

[5] Well over 90% of foreign temporary resident STEM business owners in our sample report their current visa is "for temporary work," for which the examples provided to the respondent on the survey instrument include H-1B, L-1A, and L-1B. Among other (non-STEM) college-educated foreign temporary resident business owners, about 1 in 5 indicated "other" visa type, meaning their visa was not "for temporary work" (H-1B, L-1, etc.), nor "for study or training" (F-1, J-1, H-3), nor as a dependent on someone else's visa. Cognitive interviews with college-educated foreign temporary resident workers may help to clarify how respondents categorize E-2 Treaty Investor visas ("for temporary work" versus "other"), and time spent working with an OPT extension to the F-1 visas ("for study or training" versus "for temporary work" or "other").

[6] Massachusetts Technology Cooperative, "The Global Entrepreneur in Residence Pilot Program," at http://masstech.org/innovation-institute/projects-and-initiatives/global-entrepreneur-residence-pilot-program, last accessed February 6, 2015.

[7] On February 25, 2015, DHS released a final rule extending eligibility for employment authorization to H-4 dependents (spouses and unmarried children under 21 years of age) of certain H-1B non-immigrants. For information see www.uscis.gov/working-united-states/temporary-workers/employment-authorization-certain-h-4-dependentspouses.

[8] U.S. Department of State, Bureau of Consular Affairs. "Treaty Countries," at http://travel.state.gov/content/visas/english/fees/treaty.html, last accessed 11 February 2015.

[9] On August 17, 2015, the United States Citizenship and Immigration Service released a policy memorandum affecting the issuance of the L-1B visa. This guidance can be found at: www.uscis.gov/sites/default/files/USCIS/Laws/Memoranda/2015/L1B_Memorandum_8_14_15_draft_for_FINAL_4pmAPPROVED.pdf.

[10] EB-5 visas provide conditional permanent residence for an initial two-year period. Once the required conditions have been met, EB-5 visa holders can file an I-829 petition to have the conditionality on their visa removed.

[11] U.S. Citizenship and Immigration Services, "National Interest Waiver," at www.uscis.gov/eir/visa-guide/eb-2-employment-based-second-preference/national-interest-waiver, last accessed July 24, 2015.

[12] U.S. Citizenship and Immigration Services, "Understanding the EB-2 Requirements for Exceptional Ability," at www.uscis.gov/eir/visa-guide/eb-2-employment-based-second-preference/understanding-eb-2-requirementsexceptional-ability, last accessed July 24, 2015.

[13] This survey is conducted by the U.S. Census Bureau for the NSF's National Center for Science and Engineering Statistics (NCSES).

[14] We define self-employed and business owners based on respondents' reporting their principal job sector as self-employed in a non-incorporated business, professional practice, or farm, or alternatively as self-employed or business owner in an incorporated business, professional practice, or farm.

[15] We define this group as the subset of non-incorporated self-employed individuals who report they have fewer than 10 employees (zero included), but do not directly supervise any workers and have not "recommended or initiated personnel actions such as hiring, firing, evaluating or promoting others."

[16] Here indicated by high school completion outside of the United States, including in Puerto Rico or other U.S. territories.

[17] Over half (52%) of bachelor's degrees in "earth sciences and related engineering fields" are earned in geology or "geological sciences, other," with another 15% in petroleum engineering, mining and minerals engineering, and geological and geophysical engineering. These subfields together comprise 84% of this group's STEM entrepreneurs. Other subfields in this group include atmospheric sciences and meteorology, earth sciences, and oceanography, but of these only oceanography (6.5%) is proportionally represented among STEM entrepreneurs.

[18] As previously noted, we exclude medical practitioners—specifically, all diagnosing and treating health professional occupations—from the STEM entrepreneurship outcome measure, unless the respondent listed R&D among his or her work activities.

[19] Country and region dummy variables are mutually exclusive and exhaustive, and include the following: Puerto Rico and other U.S. territories; Canada; Mexico; the United Kingdom; European EU-15, E-2 Treaty countries; NonEU European E-2 Treaty countries; Russia and non-E-2 Eastern Europe; China; Taiwan; Korea; the Philippines; Vietnam; Japan and other SSE Asia E-2 Treaty countries; India; SSE Asia non-E-2 Treaty countries; Middle East & North Africa (non-E-2); Israel; Africa E-2 Treaty countries; Sub-Saharan Africa, non-E-2; Central America & Caribbean, E-2; Argentina & Chile (E-2); Central & South America, non-E-2; Australia & New Zealand.

[20] Note that, due to the phrasing of the survey question, we know only that these individuals are "dissatisfied" or "very dissatisfied" with their job's "level of responsibility," and not whether they feel they have too much or too little.

[21] In fall 2015, DHS released two proposals updating the F-1/OPT and H-1B visa programs; these had not been finalized as of February 2016.

[22] DHS announced in its Regulatory Agenda that it is planning on releasing a proposed rule that would allow certain inventors, researchers, and entrepreneurs who establish a U.S. startup entity to be considered for admittance to the United States.

In: Foreign-Born STEM Entrepreneurship
Editor: Wayne Stevens
ISBN: 978-1-53610-270-3
© 2016 Nova Science Publishers, Inc.

Chapter 2

IMMIGRATION OF FOREIGN NATIONALS WITH SCIENCE, TECHNOLOGY, ENGINEERING, AND MATHEMATICS (STEM) DEGREES*

Ruth Ellen Wasem

SUMMARY

Although the United States remains the leading host country for international students in science, technology, engineering, or mathematics (STEM) fields, the global competition for talent has intensified. A record number of STEM graduates—both U.S. residents and foreign nationals—are entering the U.S. labor market, and there is a renewed focus on creating additional immigration pathways for foreign professional workers in STEM fields. Current law sets an annual worldwide level of 140,000 employment-based admissions, which includes the spouses and children in addition to the principal (i.e., qualifying) aliens. "STEM visa" is shorthand for an expedited immigration avenue that enables foreign nationals with graduate degrees in STEM fields to adjust to legal permanent resident (LPR) status without waiting in the queue of numerically limited LPR visas. The fundamental policy

* This is an edited, reformatted and augmented version of the Congressional Research Service Publication, CRS Report for Congress R42530, dated December 17, 2012.

question is should the United States create additional pathways for STEM graduates to remain in the United States permanently?

The number of full-time graduate students in science, engineering, and health fields who were foreign students (largely on F-1 nonimmigrant visas) grew from 91,150 in 1990 to 148,923 in 2009, with most of the increase occurring after 1999. Despite the rise in foreign student enrollment, the percentage of STEM graduate students with temporary visas in 2009 (32.7%) was comparable to 1990 (31.1%). Graduate enrollments in engineering fields have exhibited the most growth of the STEM fields in recent years. About 40,000 graduate degrees were awarded to foreign STEM students in 2009, with 10,000 of those going to Ph.D. recipients.

After completing their studies, foreign students on F-1 visas are permitted to participate in employment known as Optional Practical Training (OPT), which is temporary employment that is directly related to an F-1 student's major area of study. Generally, a foreign student may work up to 12 months in OPT status. In 2008, the Department of Homeland Security (DHS) expanded the OPT work period to 29 months for F-1 students in STEM fields.

Many F-1 visa holders (especially those who are engaged in OPT) often change their immigration status to become professional specialty workers (H-1B). Most H-1B beneficiaries are typically admitted to work in STEM occupations. In FY2010, the most recent year for which detailed data on H-1B beneficiaries (i.e., workers renewing their visas as well as newly arriving workers) are available, almost 91,000 H-1B workers were employed in computer-related occupations, and they made up 47% of all H-1B beneficiaries that year.

The H-1B visa and the OPT often provide the link for foreign students to become employment- based LPRs. In total, foreign nationals reporting STEM occupations made up 44% of all of the 676,642 LPRs who were employment-based principal immigrants during the decade of FY2000- FY2009. Of all of the LPRs reporting STEM occupations (297,668) over this decade, 52% entered as professional and skilled workers. STEM graduates seeking LPR status are likely to wait in line to obtain LPR status. Those immigrating as professional and skilled workers face wait times of many years, but those who meet the criteria of the extraordinary ability or advanced degrees preference categories have a much shorter wait.

STEM visas have gained interest in the 112th Congress, and various bills with STEM visa provisions (H.R. 399, H.R. 2161, H.R. 3146, H.R. 5893, H.R.

6412, S. 1965, S. 1986, S. 3185, S. 3192, and S. 3217) have been introduced. The House Committee on the Judiciary held two hearings on STEM and other high-skilled immigration in 2011. These issues also arose during a 2011 Senate Committee on the Judiciary hearing on the economic rationale for immigration reform. On September 20, 2012, the STEM Jobs Act of 2012 (H.R. 6429) failed to receive the necessary two-thirds vote to pass under suspension of the rules. Most recently, the House passed a revised version of the STEM Jobs Act of 2012 (H.R. 6429) on November 30, 2012.

Congress has renewed its interest in facilitating the immigration of foreign professional workers in science, technology, engineering, or mathematics (STEM) fields. The STEM workforce is seen by many as a catalyst of U.S. global economic competitiveness and is likewise considered a key element of the legislative options aimed at stimulating economic growth.[1] “STEM visa” is a shorthand for an expedited immigration avenue that enables foreign nationals with graduate degrees in STEM fields to adjust their immigration status to legal permanent residence (LPR) without waiting in the queue of numerically-limited LPR visas.[2] The fundamental policy question is: should the United States create additional pathways for STEM graduates to remain in the United States permanently? Or, are current avenues adequate?

The answer to the question lies at the nexus of education policy, labor force needs, and immigration priorities. More precisely, the key elements are: the source countries of international STEM students; the hiring choices of U.S. employers; and, the statutory limits and priorities of U.S. immigration law. This report opens by presenting a statistical portrait of foreign nationals studying STEM fields in U.S. institutions. An analysis of the current avenues foreign nationals with STEM degrees use to work in the United States temporarily and permanently follows. Discussions of the legislative history, current legislation, and major issues of debate conclude the report.

BACKGROUND

The Immigration and Nationality Act (INA) currently allocates 140,000 visas annually for economic immigrants, of which over 120,000 are allocated to three employment-based preference categories. These employment-based LPR categories are (1) persons of extraordinary ability in the arts, sciences, education, business, or athletics; outstanding professors and researchers; and certain multinational executives and managers; (2) members of the professions holding advanced degrees or persons of exceptional ability; and (3) skilled workers with at least two years training, professionals with baccalaureate degrees, and unskilled workers in occupations in which U.S. workers are in short supply.[3]

Although the United States remains the leading host country for international students in STEM fields, the global competition for talent, most notably with the European Union and Asian nations, has intensified in recent years. Concerns that these educational and career opportunities foster a "brain drain" of professionals from developing nations are infrequently expressed in the current debate. Instead, some researchers warn of a "reverse brain drain" because the United States does not allocate what they consider to be a sufficient number of visas for high-skilled immigrants to fuel the economy. According to the U.S. Department of Commerce, "growth in STEM jobs was three times as fast as growth in non-STEM jobs" over the past 10 years.[4]

Other researchers maintain that a record number of STEM graduates—both U.S. residents and foreign nationals—are entering the U.S. labor market. They express concern that foreign nationals would displace U.S. residents in the STEM fields if additional visas were allocated. Some analysts observe that the only high-skilled occupations that experienced negative wage growth in recent years were technology-related occupations (e.g., computer programmers and engineers)—occupations in which highly-educated foreign nationals cluster. Almost two-thirds of the 9.3 million people in the U.S. labor market who had STEM degrees in 2010 were employed in non-STEM occupations.[5]

Caveat

There is no generally accepted definition of what specific academic disciplines "STEM" encompasses. The National Science Foundation (NSF) studies the fields broadly and includes biological, agricultural, and environmental life sciences; computer and information sciences; mathematics and statistics; the physical sciences; psychology; the social sciences; engineering; and health fields.[6] For the purposes of tracking foreign students who study in the United States, the Department of Homeland Security's (DHS's) Immigration and Customs Enforcement (ICE) defines STEM more narrowly and does not include economics, sociology, and political science; however, ICE recently expanded its list, adding fields such as animal science, educational statistics, environmental science, nutritional science, and various specializations within psychology.[7] The National Center for Education Statistics often uses the ICE definition.[8]

TEMPORARY FOREIGN STUDENTS

International students pursuing STEM degrees in the United States are part of a long tradition of foreign students studying at U.S. institutions of higher education. Since the Immigration Act of 1924, the United States has expressly permitted foreign students to study in U.S. institutions. While their presence is generally viewed as a positive one, Congress has strengthened the reporting requirements of educational institutions with foreign students.[9] Foreign students are among the various categories of foreign nationals who are known as nonimmigrants.[10]

The most common visa for foreign students is the F-1 visa. It is tailored for international students pursuing a full-time academic education.[11] Spouses and children may accompany the F-1 visa holders on F-2 visas, but are not permitted to work. Students on F visas are generally barred from off-campus employment; however, they are permitted to work in practical training that relates to their degree program, such as paid research and teaching assistantships.[12] The Department of Homeland Security (DHS) Student and Exchange Visitor Information System (SEVIS) database reports 512,884 F-1 students were in active status (i.e., enrolled full-time or otherwise engaged in approved activities) in FY2010.[13]

Foreign Nationals Earning STEM Degrees

The number of full-time graduate students in science, engineering, and health fields who were foreign students grew from 91,150 in 1990 to 148,923 in 2009. Based on the more broadly defined NSF data, **Figure 1** shows that most of the increase occurred after 1999. Despite the rise in foreign student enrollment, the percentage of STEM graduate students with temporary visas in 2009 (32.7%) was comparable to 1990 (31.1%). Graduate enrollments in engineering fields have exhibited the most growth of STEM fields in recent years.[14] Data from the National Center for Education Statistics' Integrated Postsecondary Education Data System (IPEDS) indicate that 10,000 foreign students earned doctoral degrees in STEM fields in 2009 and almost 30,000 foreign students earned masters degrees in STEM fields in 2009.[15]

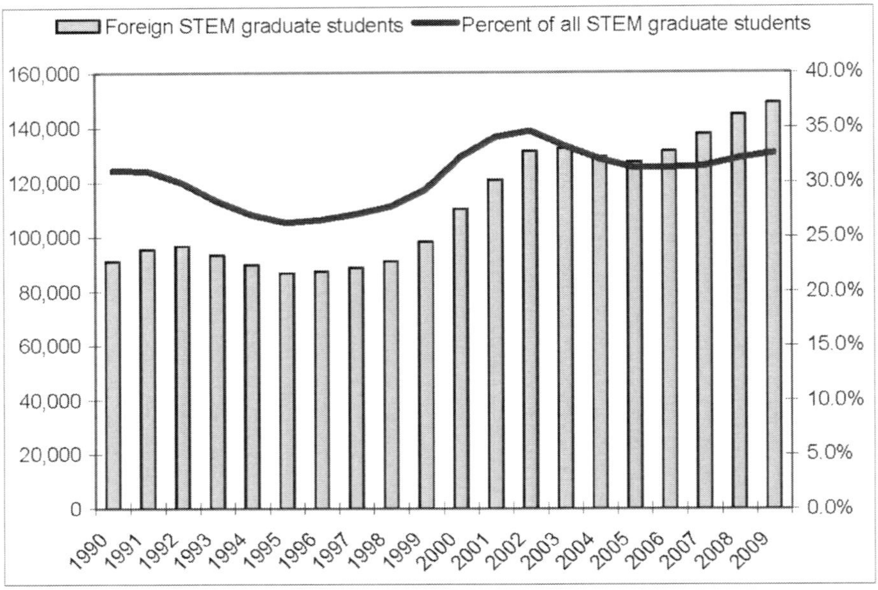

Source: National Science Foundation/National Center for Science and Engineering Statistics, NSF-NIH Survey of Graduate Students and Postdoctorates in Science and Engineering.

Figure 1. Full-Time Graduate Students with Temporary Visas in Science, Engineering, and Health Fields, 1990-2010.

A snapshot of degrees granted to foreign students by broad categories indicates that engineering is the leading field for both MA and Ph.D. recipients in 2010 (**Figure 2**). Computer science fields follow for MA degrees; while mathematics and physical sciences place second for Ph.D. degrees. Although based upon data from ICE, the STEM fields presented in **Figure 2** use the broader NSF definition of STEM, which includes economics, sociology, psychology and political science as well as agricultural and life sciences, computer sciences, mathematics and the physical sciences.

India was the top sending country for STEM graduates enrolled in masters degree programs and represented 56% of all STEM students seeking masters degrees in 2009 (**Table 1**). Indian graduate students studying engineering and computer science led in masters degrees. Students from China placed second at 15% and were more evenly distributed across STEM fields, with the exception of engineering. The remaining top eight sending countries accounted for a total of 13% of all foreign students seeking masters degrees in STEM fields in 2009, less than that of China.

In terms of doctoral degrees (**Table 2**), China sent the most STEM students in 2009, making up 35%. Chinese Ph.D. students were almost half of all foreign nationals in mathematics and physical sciences and roughly a third of all foreign nationals in the other STEM fields (except psychology and the social sciences). Doctoral degree students from India represented 16% of STEM doctorates in 2009.[16]

Table 1. Foreign Nationals Enrolled in STEM Fields, Master's Degrees 2009 Top 10 sending countries by broad categories

	Agricultural Biological Sciences	Computer Science	Economics	Engineering	Mathematics Physical Sciences	Psychology Social Sciences	Percent of Total
Total Master's	6,050	22,880	2,480	40,300	5,810	7,600	
India	2,750	16,270	140	26,290	1,210	500	56%
China	1,030	2,240	1,040	5,100	2,290	1,220	15%
Taiwan	240	520	100	1,240	200	340	3%
South Korea	140	440	140	910	280	670	3%
Nepal	160	460	50	430	110	100	2%
Saudi Arabia	70	320	40	380	90	110	1%
Canada	150	70	30	210	70	470	1%
Japan	90	70	40	130	70	520	1%
Turkey	30	130	80	380	60	190	1%
Thailand	40	250	40	330	40	90	1%

Source: Department of Homeland Security, U.S. Immigration and Customs Enforcement, Student and Exchange Visitor Information System database, special tabulations for the National Science Foundation, 2010.

Note: Although based upon data from ICE, these STEM fields use the broader NSF definition of STEM.

By broad categories

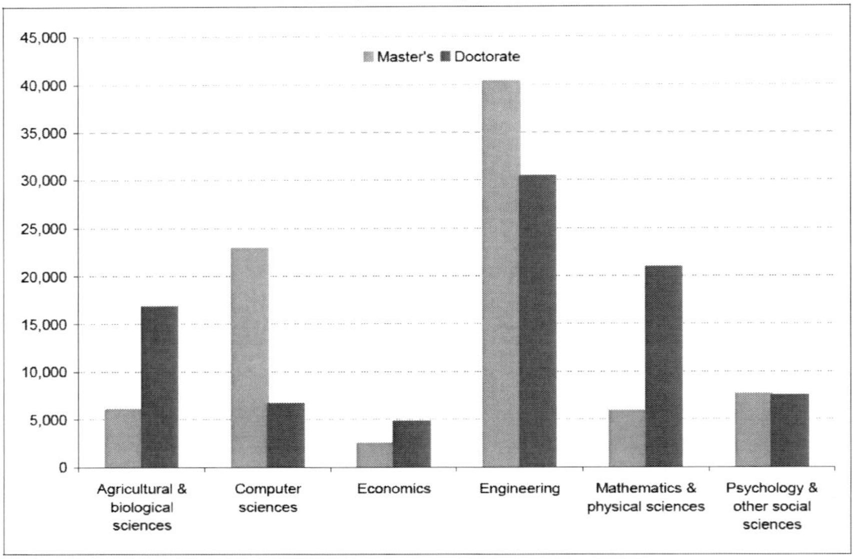

Source: Department of Homeland Security, U.S. Immigration and Customs Enforcement, Student and Exchange Visitor Information System database, special tabulations for the National Science Foundation, 2010.

Note: Although based upon data from ICE, these STEM fields use the broader NSF definition of STEM.

Figure 2. Foreign Nationals Enrolled in Advanced Degrees in STEM Fields, 2009.

Table 2. Foreign Nationals Enrolled in STEM Fields, Doctorate Degrees 2009 Top 10 sending countries by broad categories

	Agricultural Biological Sciences	Computer Science	Economics	Engineering	Mathematics Physical Sciences	Psychology Social S	Percent of Total
Total Doctorates	16,830	6,650	4,800	30,380	20,900	7,490	
China	5,550	2,410	1,300	10,570	8,680	980	35%
India	3,600	1,250	370	5,690	2,800	530	16%
South Korea	1,080	520	530	3,100	1,240	1,070	9%
Taiwan	930	220	150	1,480	700	400	4%
Turkey	250	210	280	980	440	460	3%
Iran	90	210	60	1,830	270	50	3%

	Agricultural Biological Sciences	Computer Science	Economics	Engineering	Mathematics Physical Sciences	Psychology Social	Percent of Total
Canada	460	90	80	430	380	690	2%
Japan	240	30	160	160	210	340	1%
Germany	210	70	100	150	250	210	1%
Nepal	200	40	60	160	410	50	1%

Source: Department of Homeland Security, U.S. Immigration and Customs Enforcement, Student and Exchange Visitor Information System database, special tabulations for the National Science Foundation, 2010.

Note: Although based upon data from ICE, these STEM fields use the broader NSF definition of STEM.

TEMPORARY FOREIGN WORKERS

The Immigration and Nationality Act of 1952 authorized visas for foreign nationals who would perform needed services because of their high educational attainment, technical training, specialized experience, or exceptional ability. Today, there are several temporary visa categories that enable employment-based temporary admissions for highly skilled foreign workers. They perform work that ranges from professional and skilled labor to those who work in jobs requiring extraordinary ability in the sciences, arts, education, business, or athletics.

Optional Practical Training (OPT)

After completing their undergraduate or graduate studies, F-1 foreign students are permitted to participate in employment known as Optional Practical Training (OPT), which is temporary employment that is directly related to an F-1 student's major area of study. Generally, an F-1 foreign student may work up to 12 months in OPT status. In 2008, DHS expanded the OPT work period to 29 months for F-1 students in STEM fields. To qualify for the 17-month extension, F-1 students must have received STEM degrees included on the STEM Designated Degree Program List, be employed by employers enrolled in E-

Verify,[17] and have received an initial grant of post- completion OPT related to such a degree (i.e., already approved for 12 months in OPT).[18]

According to U.S. Citizenship and Immigration Service (USCIS), the number of F-1 visa holders who are engaged in OPT was 92,465 in FY2010. They comprise 18% of all active F-1 foreign students in FY2010. STEM graduates performing OPT during their first year after graduation are included among the 82,206 F-1 students shown in **Figure 3** as "Post-Completion." In addition, there were 10,022 F-1 students who obtained the 17-month extension for OPT in STEM fields.

Temporary Professional Specialty Worker: H-1B Visas

Many F-1 visa holders (especially those who are engaged in OPT) often change their immigration status to become professional specialty workers (H-1B). Other H-1B visa holders, however, are adjusting from another immigration status or coming directly from abroad. The law sets numerical restrictions on annual admissions of the H-1B (65,000). Most H-1B workers, however, enter on visas that are exempt from the ceiling because they are returning workers or they work for universities and nonprofit research facilities that are exempt from the cap. The H-1B visas are closely associated with STEM fields, but are not limited to them.

Prospective H-1B nonimmigrants must demonstrate to USCIS that they have the requisite education and work experience for the posted positions. After the Department of Labor (DOL) approves the labor attestation, USCIS processes the petition for the H-1B nonimmigrant (assuming other immigration requirements are satisfied) for periods up to three years. An alien can stay a maximum of six years on an H-1B visa.

The H-1B labor attestation, a three-page application form, is the first step for an employer wishing to bring in an H-1B professional foreign worker. The attestation is a statement of intent rather than a documentation of actions taken.[19] In the labor attestation for an H-1B worker, the employer must attest that the firm will pay the nonimmigrant the greater of the actual wages paid to other employees in the same job or the prevailing wages for that occupation; that the firm will provide working conditions for the nonimmigrant that do not cause the working conditions of the other employees to be adversely affected; and that there is no applicable strike or lockout. The firm must provide a copy of the labor attestation to representatives of the bargaining unit or—if there is no bargaining representative—must post the labor attestation in conspicuous locations at the work site.[20]

In FY2011, there were 218,500 H-1B professional specialty worker petitions approved, down from a high of 288,000 in FY2004. Although the law sets a numerical limit of 65,000 H-1B workers each year, only initial grants are counted under the cap. As noted above, there are several categorical exceptions to the cap, most notably all H-1B nonimmigrants who work for universities and nonprofit research facilities. A provision in P.L. 108-447 also exempts up to 20,000 aliens holding a master's or higher degree from the numerical limit on H-1B visas. As **Figure 4** displays, over the past decade more H-1B workers were approved outside of the numerical cap of 65,000 than under the cap. Not all H-1B workers with approved petitions actually use the visa.

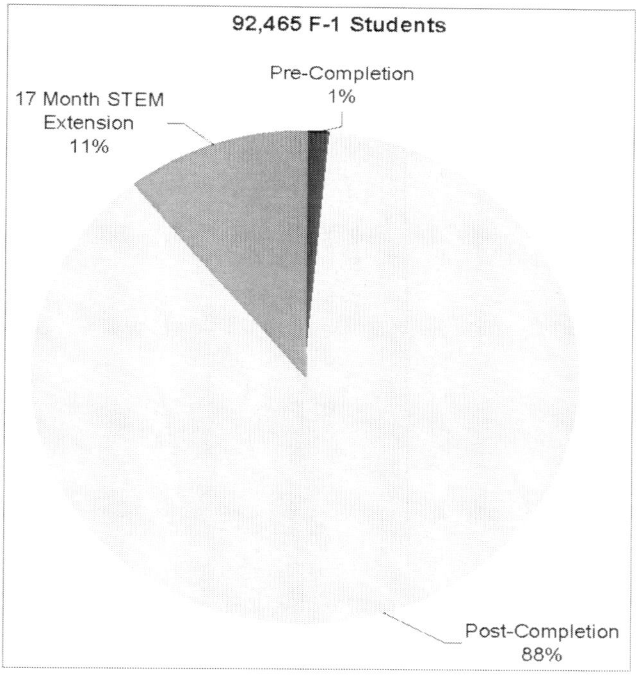

Source: CRS presentation of data from U.S. Citizenship and Immigration Services.

Notes: Post-Completion category includes those STEM graduates in the first year of OPT.

Figure 3. F-1 Foreign Students Performing Optional Practical Training in FY2010.

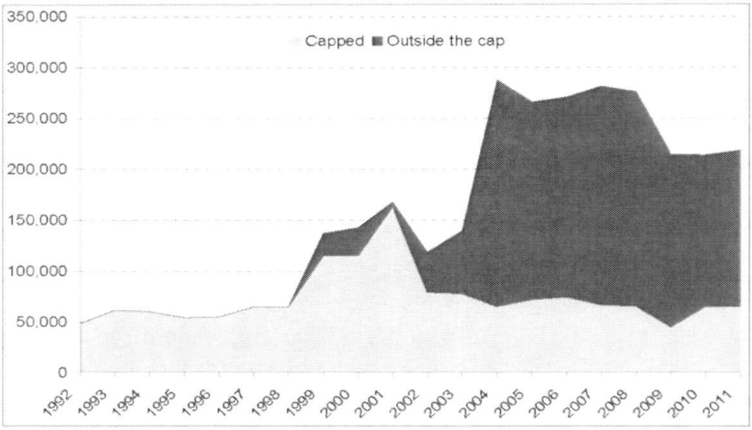

Source: CRS presentation of data from the DHS Office of Immigration Statistics (OIS) and its predecessor in the Immigration and Naturalization Service.

Notes: Congress increased the H-1B cap to 115,000 for FY1999-FY2000 and to 195,000 for FY2001-FY2003.

Figure 4. Total H-1B Petitions Approved, FY1992-FY2011.

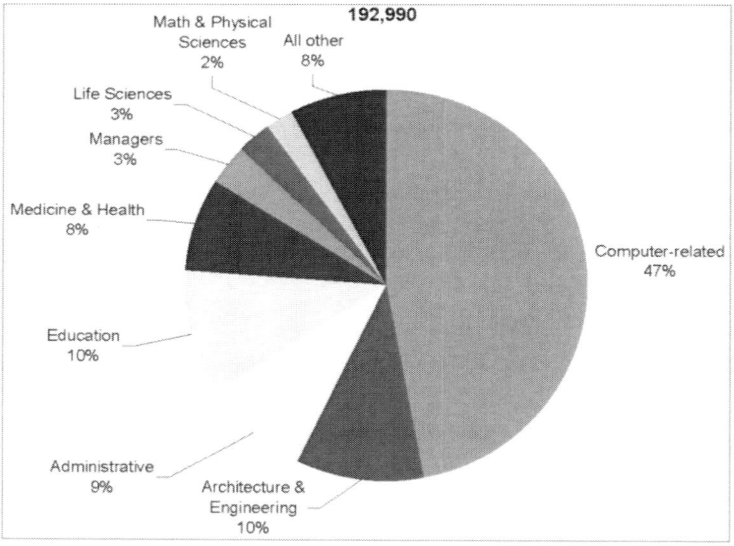

Source: CRS presentation of data from U.S. Citizenship and Immigration Services.

Note: H-1B beneficiaries include workers renewing their visas as well as newly arriving workers.

Figure 5. Occupations of H-1B Worker Beneficiaries in FY2010.

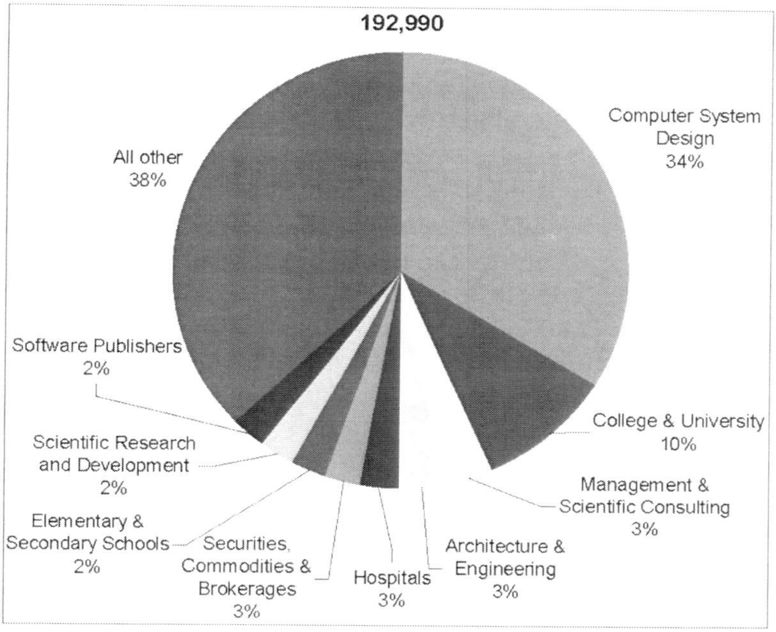

Source: CRS presentation of data from U.S. Citizenship and Immigration Services.

Note: H-1B beneficiaries include workers renewing their visas as well newly arriving workers.

Figure 6. Top Trade/Industry Sectors Hiring H-1B Worker Beneficiaries, 2010.

Over the years, a noteworthy portion of H-1B beneficiaries have worked in STEM occupations. In FY2010, the most recent year for which detailed data on H-1B beneficiaries (i.e., workers renewing their visas as well newly arriving workers) are available, almost 91,000 H-1B workers were employed in computer-related occupations, and they made up 47% of all H-1B beneficiaries that year, as **Figure 5** indicates. Architectural and engineering occupations as well as occupations in education were tied at a distant second with 10% each. Administrative occupations followed with 9%, and health and medicine occupations were 8% of the 192,990 H-1B beneficiaries. The total number of H-1B beneficiaries reported for FY2010 (192,990) and shown in **Figure 5** was less than the number of approved H-1B petitions approved that year, as depicted in **Figure 4**.[21]

In terms of the broad trade group categories in which the H-1B beneficiaries were employed in FY2010, the leading trade group was computer systems design, which employed about a third (34%) of the H-1B workers (**Figure 6**).

Colleges and universities employed 10% of the H-1B beneficiaries. Similar numbers of H-1B beneficiaries were employed in the following trade/industry sectors: management and scientific consulting; architecture and engineering; hospitals; and, securities, commodities, and brokerages. Each of these groups have hired about 3% of the H-1B beneficiaries. **Figure 6** shows that a noteworthy portion of the H-1B beneficiaries (38%) were not employed in one of the top trade/industry sectors, suggesting they are employed across the various trade/industry sectors.[22]

OTHER PROFESSIONAL SPECIALTY WORKERS: TN AND E-3 VISAS

There are two nonimmigrant visa categories quite similar to H-1B visas that are designated for temporary professional workers from specific countries. These visas are based upon specific trade agreements foreign nations have signed with the United States. Canadian and Mexican temporary professional workers may enter according to terms set by the North American Free Trade Agreement (NAFTA) on TN visas. The E-3 treaty professional visa is a temporary work visa limited to citizens of Australia.[23] Occupationally, they mirror the H-1B visa in that the foreign worker on an E-3 visa or a TN visa must be employed in a specialty occupation. They do not, however, require a labor attestation.

There are several other employment-based nonimmigrant visas for which foreign students with STEM degrees may be eligible. These nonimmigrants include cultural exchange workers on J visas; multinational executive and specialist employees on L visas; international investors on E visas; and persons with outstanding and extraordinary ability on O visas. For further information, see the Appendix, "Other High-Skilled Temporary Employment Categories."

PATHWAYS TO LEGAL PERMANENT RESIDENCE

The H-1B visa and OPT often provide the link for foreign students to become employment-based LPRs. Many anecdotal accounts tell of foreign students who are hired by U.S. firms as they are completing their programs. The employers obtain H-1B visas for the recent graduates or hire them as OPT to extend their F-1 visas. According to DHS: "This extension of the OPT period for STEM degree holders gives U.S. employers two chances to recruit these highly desirable graduates through the H-1B process, as the extension is long enough to allow for H-1B petitions to be filed in two successive fiscal years."[24] If the temporary foreign workers meet expectations, the employers may also petition for them to become LPRs through one of the employment-based immigration categories.[25]

Over 90% of 140,000 employment-based LPRs annually are adjusting to LPR status within the United States rather than newly arriving from abroad. Presumably, many of these foreign nationals had originally entered the United States as foreign students or temporary workers. Most foreign nationals seeking to qualify for a nonimmigrant visa, however, must demonstrate that they are not coming to reside permanently. Temporary workers who are H-1B visa holders are permitted to petition for a LPR visa at the same time that they file for an H-1B visa, a policy exception known as dual-intent.[26]

Permanent Employment-Based Admissions

Admissions and adjustments for LPR status are subject to a complex set of numerical limits and preference categories that give priority for admission on the basis of family relationships, needed skills, and geographic diversity. The INA establishes an annual worldwide level of 140,000 employment-based preference immigrants, and that ceiling includes the accompanying spouse and children in addition to the principal (i.e., qualifying) alien.[27] The employment-based preference categories are as follows:

- *first preference*: 40,040 for priority workers who are persons of extraordinary ability in the arts, sciences, education, business, or athletics; outstanding professors and researchers; and certain multinational executives and managers;
- *second preference*: 40,040 for members of the professions holding advanced degrees or persons of exceptional ability;
- *third preference*: 40,040 skilled workers with at least two years training, professionals with baccalaureate degrees, and unskilled workers in occupations in which U.S. workers are in short supply;
- *fourth preference*: 10,000 for special immigrants who largely consist of religious workers, certain former employees of the U.S. government, and undocumented juveniles who become wards of the court; and

- *fifth preference*: 10,000 for investors who invest at least $1 million (or less money in rural areas or areas of high unemployment) to create at least 10 new jobs.[28]

The employment-based visas made up 14.2% of the total of 1 million LPRs in FY2010.[29]

Foreign nationals with STEM degrees may qualify under several of the preference categories, depending on their talent, educational attainment, expertise, and experience. Employers who seek to hire prospective employment-based immigrants through the second and third preference categories also must petition DOL on behalf of the alien. The prospective immigrant must demonstrate that he or she meets the qualifications for the particular job as well as the preference category. If DOL determines that a labor shortage exists in the occupation for which the petition is filed, labor certification will be issued. If there is not a labor shortage in the given occupation, the employer must submit evidence of extensive recruitment efforts in order to obtain certification.[30]

From FY2000 through FY2009, 145,390 foreign nationals qualified as first preference principals (i.e., priority workers who are persons of extraordinary ability, outstanding professors and researchers or certain multinational executives and managers). Of these, 21% reported they were engaged in STEM occupations, using the broader NSF definition. As **Figure 7** depicts, almost half (49%) of the first preference principals reported executive, administrative or managerial professions.[31]

As **Figure 7** illustrates, 61% of the second preference principals reported they were engaged in STEM occupations during FY2000-FY2009. The broader NSF definition of STEM fields was used for this analysis. In other words, STEM occupations dominated among the 184,382 foreign nationals who were admitted over this decade as members of the professions holding advanced degrees or persons of exceptional ability.

STEM occupations also made up a plurality of the third preference principals (i.e., skilled workers with at least two years training, professionals with baccalaureate degrees, or unskilled workers in occupations in which U.S. workers are in short supply). From FY2000 through FY2009, 45% of the 346,870 foreign nationals admitted as third preference principals reported they worked in STEM occupations.

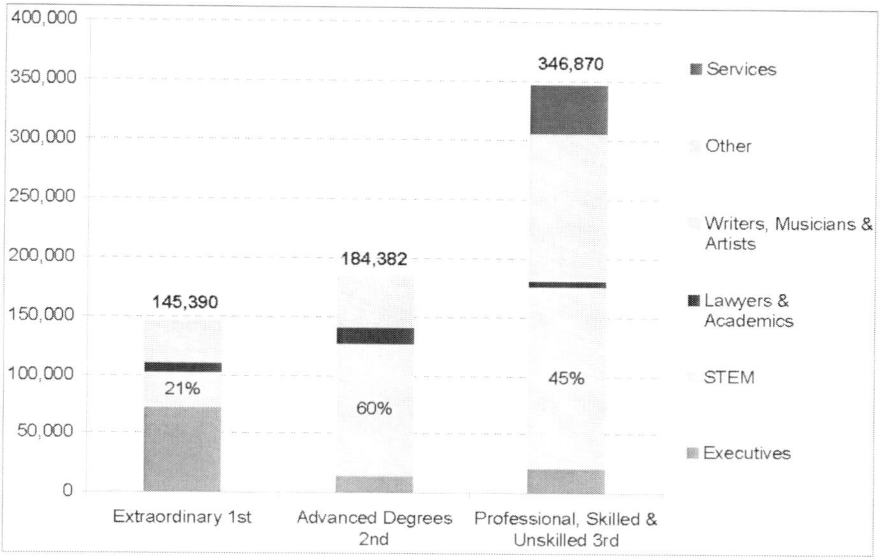

Source: CRS analysis of data from the DHS Office of Immigration Statistics and the former Immigration and Naturalization Service.

Notes: STEM includes biological, agricultural, and environmental life sciences; computer and information sciences; mathematics and statistics; the physical sciences; psychology; the social sciences; engineering; and health fields.

Figure 7. Occupations of 1st through 3rd Preference Employment-Based Principals, FY2000-FY2009.

In total, foreign nationals reporting STEM occupations made up 44% of all of the 676,642 LPRs who were employment-based principals in the first, second, and third preference categories during the decade of FY2000-FY2009. Of all of the LPRs reporting STEM occupations (297,668) over this decade, 52% entered as third preference principals.

Employment-Based I-485 Applications Pending

The volume of employment-based immigration cases pending is an important factor in the STEM debate. The USCIS maintains a system of approved employment-based I-485 applications (i.e., the Application to Register Permanent Residence or Adjust Status) that are pending, which provides another source of data on the number of approved employment-based LPRs. Known as the I-485 Inventory, these data are available by preference category and by top countries. These I- 485 data include the employment-based applicants who plan to adjust their immigration status within the United States. The prospective employment-based LPRs who would be new arrivals from abroad are not included in the I-485 inventory, because they would not need to file I-485 petitions.

As **Figure 8** shows, STEM graduates seeking LPR status are likely to wait in line to obtain LPR status, but those who meet the criteria of the extraordinary ability or advanced degrees preference categories have a much shorter wait. That there are no third preference I-485 applications filed after August 2007 in the I-485 Inventory suggests that USCIS has not been approving many since the 2007 visa retrogression pushed back the visa priority dates. These data likely understate the number of cases pending.[32]

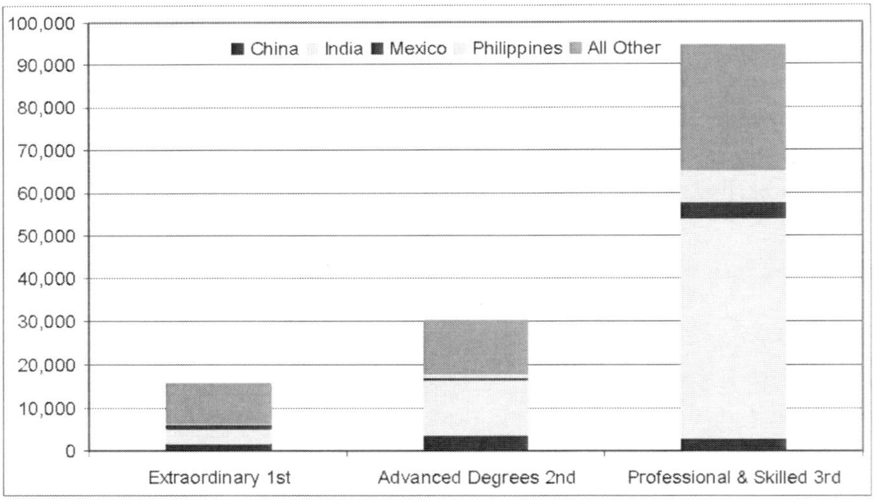

Source: USCIS I-485 Inventory of pending cases, as of January 5, 2012.

Note: There were 140,245 cases; however, no "Professional and Skilled" (3rd) preference applications filed after August 2007 are included in the I-485 Inventory thus far.

Figure 8. Employment-Based I-485 Applications Pending January 2012 by Preference and Top Countries.

LEGISLATIVE HISTORY OF STEM VISAS

Legislative proposals for STEM visas emerged during the debate over Comprehensive Immigration Reform in the mid-2000s. Various options aimed at retaining STEM graduates of American universities were folded into the legislation that was moving at the time. The approach was two-fold: (1) foster the recruitment and retention of STEM foreign students; and (2) enable foreign nationals with graduate degrees in STEM fields to adjust to LPR status without waiting in the queue of numerically limited visas. Although the ideas for STEM visas continue to be discussed, legislative action on STEM visas stalled along with the broader Comprehensive Immigration Reform efforts in 2007.

During the 109th Congress, the Senate-passed Comprehensive Immigration Reform legislation (S. 2611) was the first piece of legislation with STEM visa provisions that received action. The legislation would have exempted from numerical limits foreign nationals who had worked in the United States for three years and who had earned an advanced degree in a STEM field. More specifically, §507 of S. 2611 as passed by the Senate would have permitted foreign students on a proposed STEM student visa to adjust to LPR status. Also, §508 would have allowed an unlimited number of foreign nationals who had earned an advanced degree in a STEM field and had been working in a related field in the United States during the three years preceding to become LPRs. The 109th House of Representatives did not pass comparable legislation.

In the 110th Congress, S.Amdt. 1150 to S. 1348, the Comprehensive Immigration Reform legislation that failed a cloture vote in the Senate, as well as H.R. 1645 (Security Through Regularized Immigration and a Vibrant Economy Act of 2007, or the STRIVE Act), similarly would have created STEM visas. The legislative proposals in the 110th Congress would have established a STEM foreign student visa category and would have permitted an unlimited number of foreign nationals who had earned STEM degrees and had been working in a related field in the United States during the three years preceding to become LPRs. Foreign students seeking STEM visas would not have needed to demonstrate an intent of departing the United States upon completion of their studies. Students in this category would have also been permitted to pursue Optional Practical Training for periods of up to 24 months after completing their degree. Employers would have been required to pay foreign students the higher of the average or prevailing wage in the field of employment.

SELECTED LEGISLATION IN THE 112TH CONGRESS

There has been renewed interest in establishing STEM visas in the 112th Congress, and several bills have been introduced in both chambers. The House Committee on the Judiciary held two hearings on these issues in 2011.[33] These issues also arose during the Senate Committee on the Judiciary hearing on the economic rationale for immigration reform.[34] A variety of bills creating STEM visas have been introduced, including H.R. 399, H.R. 2161, H.R. 3146, H.R. 5893, H.R. 6412, H.R. 6429, S. 1965, S. 1986, S. 3185, S. 3192, and S. 3217.[35]

The Immigration Driving Entrepreneurship in America Act of 2011, or the IDEA Act of 2011 (H.R. 2161); the American Innovation and Education Act of 2011 (H.R. 3146); and the Science, Technology, Engineering, and Mathematics Visa Act of 2011, or STEM Visa Act of 2011 (S. 1986), while not identical, do include immigration provisions that are comparable in terms of STEM visas. The bills would amend the INA to establish a priority worker immigrant visa for an alien who has a qualifying: (1) master's or higher degree in a field of science, technology, engineering, or mathematics (STEM degree) from a U.S. institution of higher education; and (2) employment offer from a U.S. employer. S. 1986, H.R. 2161, and H.R. 3146 would also eliminate the foreign residency requirement for certain foreign students, enabling them to have dual intent.[36] The bills further would allow extensions of stay based upon a lengthy labor certification or priority adjudication for F-1 students, H-1B workers, and L intracompany transfers.

S. 1986, H.R. 2161 and H.R. 3146 would also amend the INA to impose a fee of $2,000 per petition on an employer who files an application to hire a foreign worker. Certain not-for-profit and university-based employers would be exempt from paying the fee. Funds from the fees collected would be deposited in the STEM Education and Training Account, which the bills would establish in the Treasury. S. 1986, H.R. 2161 and H.R. 3146 furthermore would require that 60% of the funds be used for a STEM scholarship program for low-income U.S. students enrolled in degree programs in science, technology, engineering, or mathematics.[37]

The Stopping Trained in America Ph.D.s From Leaving the Economy Act of 2011 or the STAPLE Act (H.R. 399) would amend the INA to exempt (1) from direct numerical limitations aliens who have earned a Ph.D. degree in a STEM field from a U.S. institution of higher education and who have an offer of employment from a U.S. employer in a field related to such degree; and (2) from H-1B visa numerical limitations aliens who have earned a Ph.D. degree in a STEM field from a U.S. institution, so long as that degree is required for the job the petitioning employer is filling.

The Startup Act of 2011[38] (S. 1965) and the Startup 2.0 Act of 2012 (H.R. 5893/S. 3217) would create a new conditional resident visa for up to 50,000 foreign students who have a master's or higher degree in a STEM field from a U.S. institution of higher education. The recipients of these new STEM visas would be granted conditional residence status contingent upon them remaining actively engaged in a STEM field for five consecutive years, after which they would be able to adjust to legal permanent resident status.

Rather than adding LPR visas for STEM graduates, the Securing the Talent America Requires for the 21st Century Act of 2012, or STAR Act (S. 3185), would eliminate the Diversity Visa Lottery and re-allocate those 55,000 Diversity Visas to a new category it would create for foreign STEM graduates who have job offers in related fields.[39] In addition, S. 3185 would amend the labor certification requirements in the INA to create a special provision in the law for employers of foreign STEM master's degree recipients, which would require the employers to engage in a competitive recruitment and selection process and determine that the alien is more qualified than any U.S. worker who applied for the job. S. 3185 would further amend the INA so that job openings in which employers are petitioning to hire foreign STEM graduates with doctoral degrees would automatically be deemed as shortage occupations that do not require certification from the Department of Labor.

The Attracting the Best and the Brightest Act of 2012 (H.R. 6412) would add 55,000 visas for foreign STEM graduates who have job offers in related fields, but it would not eliminate the Diversity Visa Lottery. Any visas unused by the foreign STEM graduates would roll over to the other employment-based visa categories. It would define STEM as a field included in the Department of Education's Classification of Instructional Programs taxonomy within the summary groups of computer and information sciences and support services, engineering, mathematics and statistics, and physical sciences. H.R. 6412 would require foreign STEM graduates to have an advanced degree from an accredited public or nonprofit university classified by the National Science Foundation as either a research institution or as otherwise excelling in STEM instruction. In provisions similar to the STEM Jobs Act of 2012 (discussed below), H.R. 6412 would amend the labor certification provisions in the INA to require employers of foreign STEM graduates to submit a job order for the position with the state workforce agency in that state, which in turn is required to post the position on its official website. However, H.R. 6412 would also require that employers offer wages to the foreign STEM graduates that meet or exceed the actual wages paid to U.S. workers with similar levels of experience.

The STEM Jobs Act of 2012 (H.R. 6429) would eliminate the Diversity Visa Lottery and re- allocate those 55,000 Diversity Visas to two new categories it would create for foreign STEM graduates who have job offers in related fields.[40] Though the STEM Jobs Act of 2012 would allocate all of the 55,000 visas to the first new category for foreign nationals who have earned a Ph.D. degree in a STEM field from a U.S. institution, any remaining visas would roll down to the second new category for foreign nationals who have earned a M.A. degree in a STEM field. Unlike H.R. 6412, however, any visas unused by the foreign STEM graduates would not roll over to the other employment-based visa categories under the STEM Jobs Act of 2012.

Similar to H.R. 6412, the STEM Jobs Act of 2012 would define STEM as a field included in the Department of Education's Classification of Instructional Programs taxonomy within the summary groups of computer and information sciences and support services, engineering, mathematics and statistics, and physical sciences. The foreign STEM graduates would have to receive degrees from a doctorate-granting university that the Carnegie Foundation for the Advancement of Teaching rates as having a very high or high level of research activity or that the National Science Foundation has classified as having research activity equivalent to those institutions classified by the Carnegie Foundation. It would amend the labor certification provisions in the INA to require employers of foreign STEM graduates to submit a job order for the position with the state workforce agency in that state, which in turn is required to post the position on its official website.

On September 20, 2012, the STEM Jobs Act of 2012 (H.R. 6429) failed to receive the necessary two-thirds vote to pass under suspension of the rules. The legislation garnered 257 yeas and 158 nays, with bipartisan support as well as bipartisan opposition.[41]

Most recently, the House passed revised version of the STEM Jobs Act of 2012 (H.R. 6429) that offers several key differences from the legislation that the House considered on September 20, 2012. The House-passed version of H.R. 6429

- would permit unused STEM visas made available in fiscal years 2013 through 2016 to be utilized in future years, under terms set by the bill;
- would eliminate a provision requiring STEM visa recipients to promise to work for at least five years in the United States or for at least five years for the petitioner in a STEM field; and,

- would eliminate a provision prohibiting universities recognized for prospective STEM visa recipients from offering incentive payments to persons based on securing foreign students.

Perhaps the most significant addition to H.R. 6429 is the expansion of the nonimmigrant visa for family members with approved LPR petitions pending, commonly known as the "V" visa.[42] The bill would allow the spouses and minor children of LPRs to live in the United States on "V" visas (without employment authorization) while they wait for their LPR visas to become available. Immediate relatives of LPRs would become eligible for the V visas after one year on the visa priority date waitlist.[43]

The House passed H.R. 6429 on November 30, 2012 by a vote of 245 Yeas and 139 Nays.[44]

POLICY DISCUSSION

The fundamental policy question for Congress is should the United States create additional pathways for STEM graduates to remain in the United States permanently? Some policymakers consider establishment of STEM visas to be a natural and positive chain of events, arguing that it would be inadvisable to educate talented young people only to make them leave to work for foreign competitors. Others maintain employment-based LPR admissions are highly skewed toward high-skilled immigrants, which already provide for generous flows of foreign nationals with STEM degrees. A corollary factor is that pulling prospective LPRs with graduate STEM degrees out of the numerically limited employment-based categories would free up visas for the other prospective LPRs waiting in the employment-based queue. The policy discussion opens with two competing perspectives on the immigration prospects of international students with STEM degrees. Four themes of debate conclude the report: assessments of U.S. labor market needs, the competing definitions of STEM fields, the role of colleges and universities as gatekeepers, and the context of broader immigration reforms.

Two Perspectives on STEM Immigration Prospects

At the crux of the legislative debate is whether expedited immigration pathways for STEM foreign graduate students would be in the national interest, and two widely cited studies of the issue are illustrative of this debate. The two studies featured below yield divergent findings, in large part because they use different research designs and study different STEM populations. This section does not purport to say that one research design is superior to another or that one target population is more meritorious than the others. Rather, it offers two examples with distinct approaches to highlight the competing perspectives. One approach is from the vantage point of the immigration caseloads in the visa categories that STEM degree holders currently use to become LPRs; it explores the question of how long would foreign nationals from China and India who have STEM degrees wait to become LPRs of the United States. The other approach is a retrospective analysis of the foreign nationals who earn doctorates in STEM fields; it addresses the issue of retention rates of foreign STEM doctorates in the United States 10 years after graduation.

STEM Graduates Face Long Waits for LPR Status

A 2011 study by Stuart Anderson of the National Foundation for American Policy (NFAP) examined prospective LPRs with approved petitions to immigrate who were waiting in the queue for a numerically limited visa. Since the official data sources do not provide a complete total of cases in the pipeline, NFAP estimated some of the missing components.[45] The NFAP analysis of the data indicate that the majority of employer-sponsored LPRs at the end of 2010 were from India and China.

NFAP's Anderson estimated that a professional worker from India who applies for a professional and skilled (third preference) visa would wait approximately 70 years to obtain LPR status.[46]

Figure 9 presents the NFAP estimates for the upper end (i.e., maximum number of years) that prospective LPRs have been waiting and might continue to wait for an LPR visa according to when they filed their petitions.

The NFAP study of the pending caseload also estimated that a professional worker from China who applies for a third preference visa would wait approximately 24 years to obtain LPR status.[47] As **Figure 9** shows, the

NFAP projected further that countries other than India and China would have wait times for a third preference of four to six years. Anderson concluded "America would lose much talent as U.S.-based businesses would need to hire or place such skilled individuals abroad, rather than invest in a green card process likely to last decades."[48]

The NFAP study indicates that foreign nationals holding advanced degrees (second preference) face far shorter wait times. Only India and China are mentioned as having a noteworthy number of second preference cases pending. Indians and Chinese seeking second preference LPR visas today would have waits of approximately six years, according to the NFAP study.[49]

NFAP's Anderson estimates that an exemption of 50,000 visas a year for foreign students with STEM graduate degrees would eliminate the backlog of cases for those holding advanced degrees (second preference) within two years. He further projects that it would ease the backlog of professional and skilled (third preference) workers, eliminating the backlog within 10 years.[50]

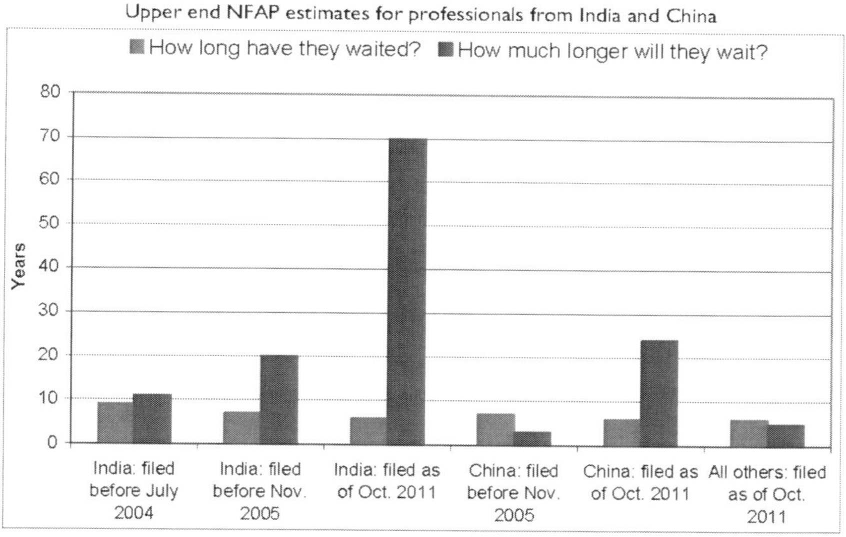

Source: *Waiting and More Waiting: America's Family and Employment-based Immigration System*, Table 2, National Foundation for American Policy (NFAP), October, 2011.

Notes: Bars depict the upper end (i.e., maximum number of years) of the NFAP estimates.

Figure 9. Projected Wait Times for Third Preference LPRs.

Stay Rates of STEM Graduates Remain High

Ongoing research that Michael G. Finn has conducted for the National Science Foundation (NSF) on foreign nationals receiving science and engineering doctorates 10 years after graduation offers a different perspective.

According to his latest published analysis, the 2009 stay rate for all foreign doctorate recipients was 64% for those graduating five years earlier and 66% for those graduating 10 years earlier. **Figure 10** has been reproduced from his report.[51]

Finn found that the 2009 stay rate of doctorate recipients on temporary resident visas at the time of graduation behaved slightly differently for different cohorts. For those graduating five years earlier, the stay rate was down slightly from that recorded two years ago. However, for those graduating 10 years earlier, the stay rate in 2009 reached an all-time high. Finn concluded that stay rates for temporary foreign nationals receiving science and engineering doctorates overall have never been higher.[52]

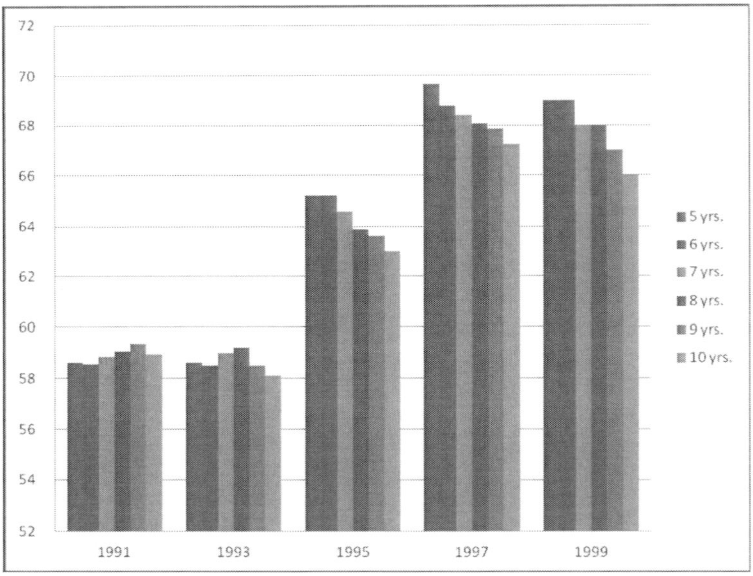

Source: reproduced from *Stay Rates of Foreign Doctorate Recipients from U.S. Universities, 2009*, by Michael G. Finn, January 2012.

Figure 10. Percentage of Foreign Nationals Receiving Science and Engineering Doctorates Who Were in the United States Five to 10 Years After Receipt of Doctorate, for Doctorates Awarded in 1991, 1993, 1995, 1997, and 1999.

Finn also studied prospective immigrants from India and China, and he reached this conclusion:

China and India are countries of special interest because they account for a large and growing share of new doctorate recipients and are subject to some restrictions not faced by most other countries when seeking permanent resident status. When comparing the history of stay rates for these two countries with that of all other countries, however, there is no apparent evidence that visa restrictions have reduced stay rates for China and India.[53]

At first glance, it seems difficult to reconcile Finn's conclusion about prospective LPRs from China and India with Anderson's estimates of how long prospective LPRs from China and India might wait before visas would be available for them. From Anderson's perspective, the wait times are too long for prospective employment-based immigrants in the professional and skilled category. From Finn's perspective, those with science and engineering doctorates have sufficient time and opportunity to become LPRs through one of the extraordinary ability or advanced degrees categories.

U.S. Labor Market Needs

A broad consensus of business, academic, and policy leaders warn that the United States is on the verge of STEM workforce shortages, which will diminish U.S. global economic competitiveness. Some analysts warn that without retaining more STEM graduates, the United States would suffer a loss of entrepreneurship, would decline in the knowledge economy, and would lose its premier place in the world of innovation. Proponents of STEM visas cite the substantial contributions of foreign-born residents in the United States to international patent creation (25.6% in 2006). "If we don't keep the skilled people in this country after they are educated in our universities and our institutions," U.S. Chamber of Commerce President Tom Donohue said, "companies have a simple choice: If we can't get them here and they go somewhere else, we send the work to where they are." The perspective of many in the business community is that skilled immigrants make jobs rather than take jobs.[54]

Some researchers maintain that creating a STEM visa would not be prudent at this time, given the soft labor market for some STEM occupations. To support this argument, they point out that the unemployment rate of 5.2% for computer and mathematical occupations was greater than the 4.7% unemployment rate of college graduates generally in 2010, and that the unemployment rate of chemists and material scientists reached its highest level in 40 years in 2011 (6.1%).[55] Earlier research from 2007 found that STEM shortages were either overstated or misunderstood. One scholar testified about research indicating that some employers already use high-skilled temporary visa categories to recruit foreign workers at below market rates. This researcher also argued that employers have undue control over the temporary foreign workers because their legal status hinges on their employment. One study found that universities in the United States actually graduate many more STEM students than are hired each year. In 2011, a policy researcher testified that the science and engineering labor force has a substantial supply and that it is best characterized as a loose (not tight) labor market.[56]

Recently, the debate over U.S. labor market needs for STEMs has centered on occupations in computer sciences. Microsoft published a report projecting the graduation rates in computer science will not keep pace with estimated job openings through 2020 in the computer science occupations that require at least a bachelor's degree.[57] The Economic Policy Institute published a report critiquing the Microsoft analysis and arguing that there is no shortage of workers in computer-related occupations.[58]

Defining/Refining STEM Fields

As noted at the onset, there is no consensus on the definition of STEM fields within academia or federal agencies. Concerns have been expressed that linking LPR status to a STEM degree without a clearly accepted definition might well have unintended consequences. For example, as a result of revisions in federal Standard Occupational Classification (SOC) in STEM occupations, there was a significant reclassification of employees. Roughly 5% of jobs in the U.S. labor force currently are considered STEM positions. The broader the STEM fields are defined, the more likely that no critical discipline would be omitted; however, a broad definition, some maintain, would also be more susceptible to gaming and abuses.[59]

Others point out that even the narrowest definition would include widely heterogeneous disciplines with widely varied labor market conditions. For example, the engineering profession, which has evolved and splintered into sub-specialties over time, was defined in 1993 as follows:

> Civil engineers had more specialized training as structural engineers, dam engineers, water- power engineers, bridge engineers; mechanical engineers as machine-design engineers, industrial engineers, motive-power engineers; electrical engineers as power and communication

> engineers (and the latter divided eventually into telegraph, telephone, radio, television, and radar engineers, whereas the power engineers divided into fossil-fuel and nuclear engineers); mining engineers as metallic-ore mining engineers and fossil-fuel mining engineers (the latter divided into coal and petroleum engineers).[60]

Today, engineering also encompasses such professions as computer engineers designing micro- chips that use light pulses and biomedical engineers researching the structure of human cells to foster tissue growth. To deal with the issue of defining STEM fields, one witness during a 2011 House hearing recommended first clarifying the policy motivations for the immigration benefit and then assigning the task of enumerating eligible fields to an agency or inter-agency work- group as an option.[61]

Confounding the discussion of what disciplines should be considered "STEM fields" is the perspective that STEM degrees are not as important as competency in a core set of STEM-based capabilities (referred to as STEM competency). There is an increasing demand for STEM competency in non-STEM occupations, which enables many people with STEM competencies to choose higher-paying career options in other fields. One report finds that out of every 100 students with a bachelor's degree, 19 graduate with a STEM degree but only eight are working in STEM occupations 10 years after graduation.[62]

Colleges and Universities as Immigration Gatekeepers

Proponents of STEM visa legislation assert that completion of a graduate degree in a STEM field at a U.S. institution is an excellent criterion for selecting LPRs. They argue that the United States should not be sending back the world's brightest students after educating them. They maintain that current policy not only fosters a reverse brain drain from our economy, but it also equips nations who compete with the United States in the global economy with graduates trained by our world-class universities. A witness at a 2011 hearing on STEM visas testified that a significant majority of foreign students—85% of Indian and Chinese students and 72% of European students—reported concerns about obtaining work visas. The witness warned that the United States was falling behind in the global competition for talent.[63]

Other policymakers question the potential consequences of linking LPR status to student visas. For example, the Chairman of the House Committee on the Judiciary warned that establishment of STEM visas would create an incentive for some schools to recruit tuition-paying foreign students with the lure of LPR status upon graduation and cited reports from Australia, where some institutions of higher education were deemed to be "selling education for visas." Another study of the unintended consequences of Australia's linkage of its skilled migration program with international students reported " the failure of some former overseas students to achieve employment outcomes that were commensurate with their qualifications; and failure to obtain skill levels that would meet Australia's skill needs." Further concerns are the possibilities that the incentive of STEM visas may lead to foreign nationals displacing qualified American students or may lead to an over-production of advanced STEM degrees. Some experts advise that institutional eligibility should be carefully defined to prevent "diploma mills" from cashing in on the STEM visas and offer options such as limiting the eligible institutions to those requiring residency at the institution or receiving research funding from NSF or the National Institutes of Health.[64]

Context of Broader Immigration Reform

The challenge in formulating an immigration system is structuring it to represent the country's values, priorities, and needs. The difficulty is designing a system that operationalizes these specific values, priorities, and needs into immigration policy. STEM visas are appealing to some because they offer an option that appears to be solely merit-based. Yet, such specification of merit itself provokes debate over what human capital, personal traits, and values prospective immigrants should bring to a country. Some maintain that the STEM visas legislation has broad enough support to be enacted without tackling the thornier issues that confound comprehensive immigration reform, such as the immigration status of unauthorized students who were brought to the United States as children by their parents.[65] Others maintain it merely begs the question by increasing legal immigration without addressing wider reforms.

APPENDIX. OTHER HIGH-SKILLED TEMPORARY EMPLOYMENT CATEGORIES

Cultural Exchange Workers: J Visas[66]

The broadest category for cultural exchange is the J visa, which includes professors and research scholars, students, foreign medical graduates, teachers, resort workers, camp counselors and au pairs who are participating in an approved exchange visitor program. The U.S. Department of State's Bureau of Educational and Cultural Affairs (BECA) is responsible for approving the cultural exchange programs. J visa holders are admitted for the period of the program. Many foreign nationals on J-1 visas are permitted to work as part of their cultural exchange program participation. The J cultural exchange visas have expanded over time from visas issued for educational, research, or scholarship purposes to visas issued for programs engaged in more mundane tasks, such as child care, resort work, or camp counseling. Today, the J visas may be issued to over a dozen subcategories of exchange visitors. Many of the J visa holders are coming to work in the United States, and some may be employed in STEM fields.

Multinational Executive and Specialist Employees: L Visas

Intracompany transferees who are executive, managerial, or have specialized knowledge and who are employed with an international firm or corporation are admitted on the L visas. The prospective L nonimmigrant must demonstrate that he or she meets the qualifications for the particular job as well as the visa category. The foreign national must have been employed by the firm for at least six months in the preceding three years in the capacity for which the transfer is sought. The alien must be employed in an executive capacity, a managerial capacity, or have specialized knowledge of the firm's product to be eligible for the L visa. STEM graduates are most likely to fulfill the specialized knowledge element. The INA does not require firms who wish to bring L intracompany transfers into the United States to demonstrate that U.S. workers will not be adversely affected in order to obtain a visa for the transferring employee.

International Investors: E Visas

Aliens who are treaty traders enter on E-1 visas, whereas those who are treaty investors enter on E-2 visas. An E-1 treaty trader visa allows a foreign national to enter the United States for the purpose of conducting "substantial trade" between the United States and the country of which the person is a citizen. An E-2 treaty investor can be any person who comes to the United States to develop and direct the operations of an enterprise in which he or she has invested, or is in the process of investing, a "substantial amount of capital." Entrepreneurial STEM graduates, in particular, may utilize the E-2 visa. Both these E-class visas require that a treaty exist between the United States and the principal foreign national's country of citizenship.[67]

Persons with Outstanding and Extraordinary Ability: O Visas

Persons with extraordinary ability in the sciences, arts, education, business, or athletics can be admitted on O visas. Generally, the O visa is reserved for the highest level of accomplishment and covers a fairly broad set of occupations and endeavors, including athletics and entertainment. Regulations implementing the O-1 visa define extraordinary ability in the field of science, education, business, or athletics as "a level of expertise indicating that the person is one of the small percentage who have arisen to the very top of the field of endeavor."[68] In FY2010, the State Department issued 8,589 O-1 visas.

End Notes

[1] For a fuller discussion of science, technology, engineering, or mathematics (STEM) fields, see CRS Report R42642, *Science, Technology, Engineering, and Mathematics (STEM) Education: A Primer*, by Heather B. Gonzalez and Jeffrey J. Kuenzi; CRS Report R42430, *America COMPETES 2010 and the FY2013 Budget*, by Heather B. Gonzalez; and CRS Report R42470, *An Analysis of STEM Education Funding at the NSF: Trends and Policy Discussion*, by Heather B. Gonzalez.

[2] For a complete discussion of the employment-based visa queue, see CRS Report R42048, *Numerical Limits on Employment-Based Immigration: Analysis of the Per-Country Ceilings*, by Ruth Ellen Wasem.

[3] The other two economic preference categories are: special immigrants who largely consist of religious workers and certain former employees of the U.S. government; and investors who invest at least $1 million (or less money in rural areas or areas of high unemployment) to create at least 10 new jobs. §203(b) of INA; 8 U.S.C. §1153.

[4] Committee on Science, Engineering, and Public Policy, *Policy Implications of International Graduate Students and Postdoctoral Scholars in the United States*, National Academies

Press, 2005; Vivek Wadhwa, Guillermina Jasso, and Ben Rissing, et al., *Intellectual Property, the Immigration Backlog, and a Reverse Brain-Drain, part III*, Duke University, New York University, Harvard Law School and the Ewing Marion Kauffman Foundation, August 2007; and, David Langdon, George McKittrick, and David Beede, et al., *STEM: Good Jobs Now and for the Future*, U.S. Department of Commerce Economics and Statistics Administration, 03-11, July 2011, http://www.esa.doc.gov/sites/ default/files/ reports/documents/stemfinalyjuly14_1.pdf.

[5] Thomas Lemieux, "The Changing Nature of Wage Inequality," *Journal of Population Economics*, vol. 21, no. 1 (January 2008); David Card and John E. DiNardo, "Skill-Biased Technological Change and Rising Wage Inequality: Some Problems and Puzzles," *Journal of Labor Economic*, vol. 20, no. 4 (2002); B. Lindsay Lowell, Hal Salzman, and Hamutal Bernstein, *Steady as She Goes? Three Generations of Students through the Science and Engineering Pipeline*, Rutgers University and the Institute for the Study of International Migration, October 2009; and, U.S. Congress, House Committee on the Judiciary, Subcommittee on Immigration Policy and Enforcement, *STEM the Tide: Should America Try to Prevent an Exodus of Foreign Graduates of U.S. Universities with Advanced Science Degrees,* testimony of Dr. B. Lindsey Lowell, 112th Cong., 1st sess., October 5, 2011; and David Langdon, George McKittrick, and David Beede, et al., *STEM: Good Jobs Now and for the Future*, U.S. Department of Commerce Economics and Statistics Administration, 03-11, July 2011, http://www.esa.doc.gov/sites/default/files/reports/documents/ stemfinalyjuly14_1.pdf.

[6] Peter Einaudi, *Two Decades of Increasing Diversity More than Doubled the Number of Minority Graduate Students in Science and Engineering*, National Science Foundation, NSF 11-319, July 2011, http://www.nsf.gov/statistics/infbrief/ nsf11319/nsf11319.pdf.

[7] For a detailed list of the Immigration and Custom Enforcement STEM-designated degree programs, see http://www.ice.gov/doclib/sevis/pdf/stem-list-2011.pdf.

[8] National Center for Education Statistics, *Postsecondary Awards in Science, Technology, Engineering, and Mathematics, by State: 2001 and 2009*, U.S. Department of Education, April 2011, http://nces.ed.gov/pubs2011/ 2011226.pdf.

[9] CRS Report RL32188, *Monitoring Foreign Students in the United States: The Student and Exchange Visitor Information System (SEVIS)*, by Alison Siskin; and CRS Report RL31146, *Foreign Students in the United States: Policies and Legislation*, by Chad C. Haddal.

[10] Nonimmigrants are admitted for a designated period of time and a specific purpose.There are 24 major nonimmigrant visa categories, which are commonly referred to by the letter and numeral that denotes their subsection in the INA; for example, B-2 tourists, E-2 treaty investors, F-1 foreign students, or H-1B temporary professional workers. CRS Report RL31381, *U.S. Immigration Policy on Temporary Admissions*, by Ruth Ellen Wasem.

[11] Those students who wish to pursue a non-academic (e.g., vocational) course of study apply for an M visa. Their spouses and children may accompany them as M-2 nonimmigrants. Some graduate students use the J visa for cultural exchange, which covers educational, research or scholarship purposes.

[12] 8 C.F.R. 214.2(f)(15)(i).

[13] SEVIS is the foreign student monitoring database. For background on SEVIS, see CRS Report RL32188, *Monitoring Foreign Students in the United States: The Student and Exchange Visitor Information System (SEVIS)*, by Alison Siskin.

[14] Until the 2000s, the enrollment of U.S. citizens in graduate science and engineering programs had not kept pace with that of foreign students in these programs. NSF researchers cite the growth in the numbers of U.S. citizens and permanent residents pursuing graduate-level study in STEM fields in recent years as the reason the percentage of foreign STEM students has remained level. Peter Einaudi, *Two Decades of Increasing Diversity More than Doubled the Number of Minority Graduate Students in Science and Engineering*, National Science Foundation, NSF 11-319, July 2011, http://www.nsf.gov/statistics/infbriefnsf11319/ nsf11319.pdf.

[15] To access the IPEDS data, go to http://nces.ed.gov/ipeds/ on the National Center for Education Statistics website.

[16] Although based upon data from ICE, the STEM fields presented in **Figure 2**, **Table 1**, and **Table 2** use the broader NSF definition of STEM. Joan Burrelli, *Foreign Science and Engineering Students in the United States,* National Science Foundation, NSF 10-324, July 2010, http://www.nsf.gov/statistics/infbrief/nsf10324/.

[17] E-Verify is an electronic employment eligibility verification program that U.S. employers voluntarily use to confirm the new hires' employment authorization through Social Security Administration and, if necessary, DHS databases. CRS Report R40446, *Electronic Employment Eligibility Verification*, by Andorra Bruno.

[18] 8 C.F.R. 214.2(f)(10).

[19] Attestation was part of a compromise package on H-1B visas that included annual numerical limits in the Immigration Act of 1990 (P.L. 101-649). See CRS Report RL30498, *Immigration: Legislative Issues on Nonimmigrant Professional Specialty (H-1B) Workers*, by Ruth Ellen Wasem.

[20] INA §212(n); 8 C.F.R. §214.2(h)(4). For a further discussion of labor attestations, see CRS Report RL30498, *Immigration: Legislative Issues on Nonimmigrant Professional Specialty (H-1B) Workers*, by Ruth Ellen Wasem.

[21] U.S. Citizenship and Immigration Services, *Characteristics of H-1B Specialty Occupation Workers, Fiscal Year 2010 Annual Report*, Department of Homeland Security, August 4, 2011.

[22] Ibid.

[23] §501 of P.L. 109-13, the Emergency Supplemental Appropriations Act for Defense, the Global War on Terror, and Tsunami Relief, 2005.

[24] U.S. Citizenship and Immigration Services, *Extension of Post-Completion Optional Practical Training (OPT) and F- 1 Status for Eligible Students under the H-1B Cap-Gap Regulations*, U.S. Department of Homeland Security, April 2, 2010.

[25] Not all companies, however, seek to convert H-1B employees to LPR status. Research by Professor Ron Hira of the Rochester Institute of Technology indicates that many of the largest users of the H-1B visa sponsor few, if any, of their H-1Bs for permanent residency. U.S. Congress, Senate Committee on the Judiciary, Subcommittee on Immigration, Refugees and Border Security, *The Economic Imperative for Enacting Immigration Reform*, 112th Cong., 1st sess., July 26, 2011.

[26] The other categories permitted dual intent are intracompany transfers employed with international firms who enter on L visas and foreign nationals with V visas for family-related nonimmigrant. §214(b) of the INA; 8 U.S.C. §1184(b).

[27] Numbers not used in one category roll down to the next category. CRS Report RL32235, *U.S. Immigration Policy on Permanent Admissions*, by Ruth Ellen Wasem.

[28] For examples of legislation in the 112th Congress that would amend the fifth preference to foster STEM graduates who are entrepreneurs, see H.R. 1114 and S. 565.

[29] Randall Monger and James Yankay, *U.S. Legal Permanent Residents: 2010*, U.S. Department of Homeland Security Office of Immigration Statistics, Annual Flow Report, March 2011, http://www.dhs.gov/xlibrary/assets/statistics/ publications/lpr_fr_2010.pdf.

[30] See CRS Report RL33977, *Immigration of Foreign Workers: Labor Market Tests and Protections*, by Ruth Ellen Wasem.

[31] CRS analyzed the U.S. Department of Homeland Security Office of Immigration Statistics LPR data for fiscal years 2000 through 2009. There were a total of 10.3 million LPRs over the decade, of whom 69% reported an occupation. The occupation data analyzed in this report are the subset of 676,642 principal immigrants who received employment 1st, 2nd, or 3rd preference LPR status from fiscal years 2000 through 2009.

[32] A substantial visa retrogression occurred in July 2007. The *Visa Bulletin for July 2007* listed the visa priority dates as "current" for the employment-based preferences (except for the unskilled other worker category). On July 2, 2007, however, the State Department issued an

Update to July Visa Availability that retrogressed the dates to "unavailable." The State Department offered the following explanation: "The sudden backlog reduction efforts by Citizenship and Immigration Services Offices during the past month have resulted in the use of almost 60,000 Employment numbers.... Effective Monday July 2, 2007 there will be no further authorizations in response to requests for Employment-based preference cases." U.S. Department of State, Bureau of Consular Affairs, *Visa Bulletin,* No. 108, is available at http://travel.state.gov/visa/frvi/bulletin/bulletin_3266.html.

[33] U.S. Congress, House Committee on the Judiciary, Subcommittee on Immigration Policy and Enforcement, *H-1B Visas: Designing a Program To Meet the Needs of the U.S. Economy and U.S. Workers*, 112th Cong., 1st sess., March 31, 2011; and U.S. Congress, House Committee on the Judiciary, Subcommittee on Immigration Policy and Enforcement, *STEM the Tide: Should America Try to Prevent an Exodus of Foreign Graduates of U.S. Universities with Advanced Science Degrees*, 112th Cong., 1st sess., October 5, 2011.

[34] U.S. Congress, Senate Committee on the Judiciary, Subcommittee on Immigration, Refugees and Border Security, *The Economic Imperative for Enacting Immigration Reform*, 112th Cong., 1st sess., July 26, 2011

[35] For a discussion of H.R. 3012, which would address high-skilled immigration by lifting the employment-based country ceilings, see CRS Report R42048, *Numerical Limits on Employment-Based Immigration: Analysis of the Per- Country Ceilings*, by Ruth Ellen Wasem.

[36] Dual intent refers to exceptions to §214(b) of the INA that permit foreign nationals to have pending petitions for both nonimmigrant and immigrant visas.

[37] For information on similar fees, see CRS Report RL31973, *Programs Funded by the H-1B Visa Education and Training Fee, and Labor Market Conditions for Information Technology (IT) Workers*, by Linda Levine and Blake Alan Naughton.

[38] Not to be confused with the StartUp Visa Act of 2011 ((H.R. 1114/S. 565) which would amend the INA to establish an employment-based, conditional immigrant visa (StartUp visa) for a sponsored alien entrepreneur: (1) with required amounts of financial backing from a qualifying investor, government entity, or venture capitalist; and (2) whose commercial activities will generate required levels of employment, revenue, or capital investment.

[39] For background on the Diversity Visa Lottery, see CRS Report R41747, *Diversity Immigrant Visa Lottery Issues*, by Ruth Ellen Wasem.

[40] The STEM Jobs Act of 2012 is posted on the website of the House Rules Committee at http://docs.house.gov/.

[41] Roll Call 590, available at http://clerk.house.gov/evs/2012/roll590.xml.

[42] For further background on the "V" visa, see CRS Report RL31381, *U.S. Immigration Policy on Temporary Admissions*, by Ruth Ellen Wasem, p.8.

[43] For a discussion of the family-based backlogs, see CRS Report RL32235, *U.S. Immigration Policy on Permanent Admissions*, by Ruth Ellen Wasem, pp. 11-13.

[44] Roll Call 613, available at http://clerk.house.gov/evs/2012/roll613.xml.

[45] For a full discussion of the data problems in estimating backlogs of approved petitions, see CRS Report R42048, *Numerical Limits on Employment-Based Immigration: Analysis of the Per-Country Ceilings*, by Ruth Ellen Wasem.

[46] The study attempted to address the unknown cases pending who were waiting in the queue for a numerically limited visa and estimated that an additional 150,000 prospective LPRs from India were waiting in the professional and skilled worker preference queue. That estimate yielded a total of 210,000 Indians in the third preference caseload. Stuart Anderson, *Waiting and More Waiting: America's Family and Employment-based Immigration System,* National Foundation for American Policy, NFAP Policy Brief, October 2011. (Hereinafter Anderson, *Waiting and More Waiting: America's Family and Employment-based Immigration System,* NFAP Policy Brief, October 2011.)

[47] The study assumed that 55,000 Chinese were already in the third preference caseload. Anderson, *Waiting and More Waiting: America's Family and Employment-based Immigration System*, NFAP Policy Brief, October 2011.

[48] Anderson, *Waiting and More Waiting: America's Family and Employment-based Immigration System*, NFAP Policy Brief, October 2011.

[49] Anderson, *Waiting and More Waiting: America's Family and Employment-based Immigration System*, NFAP Policy Brief, October 2011.

[50] Anderson, *Waiting and More Waiting: America's Family and Employment-based Immigration System*, NFAP Policy Brief, October 2011.Anderson also factors in the calculation of removing the per-country ceilings on employment- based immigrants. For more on this option, see CRS Report R42048, *Numerical Limits on Employment-Based Immigration: Analysis of the Per-Country Ceilings*, by Ruth Ellen Wasem.

[51] Michael G. Finn, *Stay Rates of Foreign Doctorate Recipients from U.S. Universities, 2009*, Division of Science Resources Statistics of the National Science Foundation by Oak Ridge Institute for Science and Education through an interagency agreement with Department of Energy, January 2012, http://orise.orau.gov/files/sep/stay-rates-foreign- doctorate-recipients-2009.pdf. (Hereinafter Finn, *Stay Rates of Foreign Doctorate Recipients from U.S. Universities, 2009*.)

[52] Finn, *Stay Rates of Foreign Doctorate Recipients from U.S. Universities, 2009*.

[53] "While stay rates of India and China decline with years since graduation, this decline has been very slight and smaller than the corresponding decline in stay rates observed by all other countries combined." Michael G. Finn, *Stay Rates of Foreign Doctorate Recipients from U.S. Universities, 2009*, Division of Science Resources Statistics of the National Science Foundation by Oak Ridge Institute for Science and Education through an interagency agreement with Department of Energy, January 2012, http://orise.orau.gov/ files/sep/stay-rates-foreign-doctorate-recipients-2009.pdf.

[54] National Academy of Sciences, National Academy of Engineering, and Institute of Medicine, *Rising Above the Gathering Storm: Energizing and Employing America for a Brighter Economic Future*, Committee on Prospering in the Global Economy of the 21st Century, 2007; Brookings Institution and George Mason University, *Immigration Policy: Highly Skilled Workers and U.S. Competitiveness and Innovation*, Forum hosted by the Brookings Center for Technology Innovation and the George Mason Center for Science and Technology Policy, February 7, 2011; Vivek Wadhwa, Guillermina Jasso, and Ben Rissing, et al., *Intellectual Property, the Immigration Backlog, and a Reverse Brain-Drain, part III*, Duke University, New York University, Harvard Law School and the Ewing Marion Kauffman Foundation, August 2007; U.S. Congress, House Committee on the Judiciary, Subcommittee on Immigration Policy and Enforcement, *H-1B Visas: Designing a Program To Meet the Needs of the U.S. Economy and U.S. Workers*, 112th Cong., 1st sess., March 31, 2011; U.S. Congress, House Committee on the Judiciary, Subcommittee on Immigration Policy and Enforcement, *STEM the Tide: Should America Try to Prevent an Exodus of Foreign Graduates of U.S. Universities with Advanced Science Degrees?*, 112th Cong., 1st sess., October 5, 2011; and, Mallie Jane Kim, "Chamber of Commerce, Bloomberg Push Immigration Reform," *U.S. News & World Report*, September 28, 2011.

[55] U.S. Congress, Senate Committee on the Judiciary, Subcommittee on Immigration, Refugees and Border Security, *The Economic Imperative for Enacting Immigration Reform*, answers to questions for the record, witness Professor Ron Hira, 112th Cong., 1st sess., July 26, 2011; and Rudy M. Baum, "Unemployment Data Worst In 40 Years," *Chemical and Engineering News*, March 21, 2012.

[56] Richard Freeman, "The Market for Scientists and Engineers," *NBER Reporter*, no. 3 (Summer 2007); Rudy M. Baum, "Unemployment Data Worst In 40 Years," *Chemical and Engineering News*, March 21, 2012; U.S. Congress, Senate Committee on the Judiciary, Subcommittee on Immigration, Refugees and Border Security, *The Economic Imperative for Enacting Immigration Reform*, answers to questions for the record, witness Professor Ron Hira, 112th

Cong., 1st sess., July 26, 2011U.S. Congress, House Committee on the Judiciary, Subcommittee on Immigration Policy and Enforcement, *H-1B Visas: Designing a Program To Meet the Needs of the U.S. Economy and U.S. Workers*, testimony of Professor Ron Hira, 112th Cong., 1st sess., March 31, 2011; and, U.S. Congress, House Committee on the Judiciary, Subcommittee on Immigration Policy and Enforcement, *STEM the Tide: Should America Try to Prevent an Exodus of Foreign Graduates of U.S. Universities with Advanced Science Degrees,* testimony of Dr. B. Lindsey Lowell, 112th Cong., 1st sess., October 5, 2011.

[57] Microsoft Corporation, *A National Talent Strategy" Ideas for Securing U.S. Competitiveness and Economic Growth* (2012), http://www.microsoft.com/en-us/news/download/presskits/citizenship/MSNTS.pdf.

[58] Daniel Costa, *STEM labor shortages?: Microsoft Report Distorts Reality About Computing Occupations*, Economic Policy Institute, November 19, 2012, http://www.epi.org/publication/pm195-stem-labor-shortages-microsoft-report- distorts/.

[59] "The large decline in the numbers of people who used to be grouped in this field is that the definition of "computer scientists and systems analysts" has become much more precise, and many of the people who might have been included in the more broadly defined profession during the 1990s are now grouped elsewhere under numerous new occupational titles for specialized types of information technology people, such as software engineers, network analysts, and database administrators." Richard Ellis, *Effects of Recent Revisions in Federal Standard Occupational Classification (SOC) Categories on Counts of the Employment of STEM Professionals*, Commission on Professionals in Science and Technology, May 2, 2007, http://www.cpst.org/NewPubs.cfm; Robert D. Atkinson, "Why the Current Education Reform Strategy Won't Work," *Issues in Science and Technology*, Spring 2012, pp. 29-36; and U.S. Congress, House Committee on the Judiciary, Subcommittee on Immigration Policy and Enforcement, *STEM the Tide: Should America Try to Prevent an Exodus of Foreign Graduates of U.S. Universities with Advanced Science Degrees?*, testimony of Barmack Nassirian, 112th Cong., 1st sess., October 5, 2011.

[60] Joseph W. Barker, *McGraw-Hill Encyclopedia of Engineering*, ed. McGraw-Hill, 2nd ed. (1993), pp. 409-410.

[61] U.S. Congress, House Committee on the Judiciary, Subcommittee on Immigration Policy and Enforcement, *STEM the Tide: Should America Try to Prevent an Exodus of Foreign Graduates of U.S. Universities with Advanced Science Degrees?*, testimony of Barmack Nassirian, 112th Cong., 1st sess., October 5, 2011.

[62] "At the graduate degree level of educational attainment, it is clear that both Managerial and Professional occupations and Healthcare Professional occupations pay substantially more than STEM." Anthony P. Carnevale, Nicole Smith, and Michelle Melton, *STEM: Science, Technology, Engineering, Mathematics*, Georgetown University Center on Education and the Workforce, October 2011, V=http://cew.georgetown.edu/stem/.

[63] U.S. Congress, Joint Economic Committee, *Fueling Local Economies: Research, Innovation and Jobs*, testimony of Robert Litan, 111th Cong., June 29, 2010; U.S. Congress, House Committee on the Judiciary, Subcommittee on Immigration Policy and Enforcement, *STEM the Tide: Should America Try to Prevent an Exodus of Foreign Graduates of U.S. Universities with Advanced Science Degrees?*, testimony of Vivek Wadhwa, 112th Cong., 1st sess., October 5, 2011.

[64] U.S. Congress, House Committee on the Judiciary, Subcommittee on Immigration Policy and Enforcement, *STEM the Tide: Should America Try to Prevent an Exodus of Foreign Graduates of U.S. Universities with Advanced Science Degrees?*, 112th Cong., 1st sess., October 5, 2011; Elsa Koleth, *Overseas Students: Immigration Policy Changes 1997– May 2010*, Parliament of Australia, Parliamentary Library Background Note, June 18, 2010, http://www.aph.gov.au/ binaries/library/pubs/bn/sp/overseasstudents.pdf.

[65] For a complete analysis of these immigration issues, see CRS Report R41704, *Overview of Immigration Issues in the 112th Congress*, by Ruth Ellen Wasem; and CRS Report R42036, *Immigration Legislation and Issues in the 112th Congress*, coordinated by Andorra Bruno.

[66] The Q visa is also an employment-oriented cultural exchange program. Its stated purpose is to provide practical training and employment as well as to share the history, culture, and traditions of the United States and of the exchange visitor's home country. While STEM graduates are not typically recruited in Q exchange programs, they are not precluded.

[67] See CRS Report RL33844, *Foreign Investor Visas: Policies and Issues*.

[68] 8 C.F.R. '214.2(o)(3)(ii).

INDEX

D

E

F

G

H

I

J

K

L

M

N

O

P

Q

R

S

Y